GOALS IN
SPACE

GOALS IN
SPACE

AMERICAN VALUES AND THE FUTURE
OF TECHNOLOGY

WILLIAM SIMS BAINBRIDGE

State University of New York Press

Published by
State University of New York Press, Albany

© 1991   State University of New York

For information, address State University of New York
Press, State University Plaza, Albany, N.Y. 12246

Production by Marilyn Semerad
Marketing by Fran Keneston

Library of Congress Cataloging-in-Publication Data

Bainbridge, William Sims.
    Goals in space : American values and the future of technology /
William Sims Bainbridge.
        p. cm.
    Includes bibliographical references and index.
    ISBN 0-7914-0614-8 (cloth) . — ISBN 0-7914-0615-6 (paper)
    1. Astronautics—Social aspects—United States. 2. College
. students—United States—Attitudes. I. Title.
    TL789.8.U5B33   1991
    333.9'4'0973—dc20
                                                            90-9901
                                                            CIP

10 9 8 7 6 5 4 3 2 1

to Wilma Alice Bainbridge
for whom everything is the future

CONTENTS

CHAPTER ONE

INTRODUCTION

In the final third of the twentieth century, American culture assimilated the possibility of spaceflight and grappled with the question of what goals to seek beyond the Earth. A revolutionary technological social movement had maneuvered the major nations into developing a near-Earth spacefaring capability, but the social conditions required for great further steps seemed absent. Therefore, it is crucial to survey the culture's evaluation of spaceflight, to learn if it understands the practical benefits of the limited space development already achieved and if it imagines a universe of more radical possibilities that a renewed thrust forward might gain.

This book delineates the values spaceflight holds for American culture and identifies more than a hundred specific goals in space that Americans find plausible. Based on approximately 4,000 questionnaires, it reports both the precise words through which our culture discusses space and the statistical correlations that link the specific ideas in the public mind. Beyond the practical benefits that exploitation of near-Earth orbit has given our economy and communications system, it probes for idealistic and long-term goals that have begun to have meaning for members of our society. Utilitarian motives may keep the space program aloft, but there is serious question whether they alone can drive it to new heights of achievement.

Although a growing library of books and articles asserts that deep space will soon return great rewards, there remains great doubt whether the Earth would benefit economically from exploiting space beyond geosynchronous orbit. Were this the nineteenth century, it might make sense to talk about the value of iron in the asteroids. But in the modern world of synthetic materials it is hard to imagine that any ordinary raw material would be worth bringing back to Earth from elsewhere in the solar system. Even precious gems are routinely synthesized. Occasionally, scarcities force uncomfortable choices, for example the dilemma the United States faces in getting its chromium either from the Soviet Union or from South Africa. But the big problems are plentiful energy, water, and clean air. Spaceflight near the Earth can make major contributions here, but for the solution of terrestrial economic difficulties, Mars does not matter.

Therefore, the conquest of deep space may have to rely upon non-economic motives, even upon irrationalities that may lurk deep within the human spirit. If so, something is wrong with that general theory of human history known as *technological determinism* (Ogburn 1922; L. White 1959). This is the view that technology is the engine of history, that technical inventions produce all other significant developments in society, and that technology is essentially self-generating. I cannot here do justice to this broad perspective, but with respect to spaceflight it might say that interplanetary rockets were bound to develop once the necessary support technologies were in place and enough scientific research had been performed. However, social scientists have demonstrated that societies do not automatically invest in all the technologies potentially available to them, and some very special social mechanism is necessary if the particular development in question does not have rather immediate economic payoffs (Schmookler 1966; Merton 1970; Simon 1971).

An alternative analysis might begin with Thomas Kuhn's model of scientific revolutions (Kuhn 1959, 1962; cf. Gutting 1980). Kuhn had studied the Copernican revolution in astronomy that displaced the Earth from the center of things, and he became convinced that such scientific revolutions were important episodes in intellectual history. In contrast, most scientific work is decidedly unrevolutionary, adding tiny bits of knowledge to already established conceptual frameworks. This Kuhn called *normal science*, work based on prior achievements, follow-

ing a well-established *paradigm*, a tradition of accepted methodological practice and agendas for research. *Scientific revolutions*,
in contrast, overturn old paradigms and establish entirely new
perspectives and agendas.

Whether Kuhn is right that scientific revolutions are sometimes necessary and normal science alone cannot achieve all
possible progress, the distinction surely clarifies some episodes
in scientific history and can also be applied to technology. I
believe that spaceflight was the result of a radical social movement, seeking the transcendent goals of interplanetary exploration and colonization, rather than of any more mundane process. Before the spaceflight movement demonstrated the
practicality of the multi-stage, liquid-fuel rocket, travel into
space was an impractical fantasy. Afterward, exploitation of
Earth orbit was an integral part of advanced technological culture. But the standard industrial, financial, and governmental
institutions did not invest in rocket development to better
achieve their conventional tasks; rather, they were manipulated
into the investment by the spaceflight movement.

THE ORIGINS OF SPACEFLIGHT

Fifteen years ago I wrote about the radical social movement
that achieved the measure of spaceflight so far gained by
humans (Bainbridge 1976). In four great nations, tiny spaceflight clubs emerged, stimulated by the wild dreams that had
been developed in science fiction and scientific popularizations.
The first and most important was the German club, the Society
for Space Travel, founded in 1927 (Ley 1969). The American
Interplanetary Society was born at the birthday party of a science fiction editor in 1930. The Russian Group for the Study of
Reactive Propulsion came next, in 1931, followed by the British
Interplanetary Society in 1933. These organizations were small,
lonely groups of enthusiasts with no official support from government, corporations, or scientific institutes.

In the early 1930s, as the Society for Space Travel was about to
go out of business for lack of money, it was able to convince the

German Army to develop the liquid-fuel rocket as a substitute for long-range artillery. A key factor was that the treaties which ended World War I limited German heavy guns but said nothing about rockets. However, military solid-fuel rockets had long existed, and their potential for improvement was clear. When the German branch of the spaceflight movement forged its alliance with the army, the partnership seemed good for both sides.

There is some historical question who, exactly, was responsible for this marriage (Winter 1983). An army engineering team was sent in search of rocket engineers, and the Society was hunting for a patron. When they found each other, it is hard to apportion responsibility for this success. However, it is clear that the society was an essential ingredient in the development of spaceflight technology in Germany. Had it not conducted many successful test firings of liquid-fuel rocket engines, this technology might not have seemed practical to the military engineers. And the crucial contribution the society made to the subsequent military program was personnel, in the form of Wernher von Braun and several other former members who had been inspired by the society's dreams of spaceflight and learned much from it about the principles of rocket propulsion.

Under von Braun's direction, a growing team of rocket engineers developed a series of projects that advanced spaceflight technology, including the A-4 (V-2) rocket, which was really a prototype spaceship. Employed during the last phase of the war, the V-2 was not a cost-effective weapon. It was far too small to carry a nuclear warhead, and the German nuclear program was fortunately very far behind that of the United States. Although the V-2 could have carried extremely deadly nerve gas, apparently no serious thought was ever given to this possibility, and its standard warhead was a bit less than 1 ton of high explosive.

I have suggested a sociological model of strategic interaction that describes the social process by which spaceflight advanced (Bainbridge 1976). Small groups of men, often dedicated in early adolescence to spaceflight, manipulated political situations to get their pet projects funded. Unable to get anyone to invest in spaceflight for its own sake, they often discovered a potential patron—in this case the German Army—who was locked in competition with an opponent and not doing very well in the struggle at the moment. The space leader would sell a project to this powerful patron as a partial solution for the patron's problem and use the resources gained to advance the cause of spaceflight. Although the German Army saw a potential in liquid-fuel rockets, as a way of outflanking the allies technologically and getting out from under the restrictions inflicted by its defeat in World War I, it had many choices before it and invested in this particular possibility only because von Braun and other members of the Society made their case effectively.

Walter A. McDougall (1985), argues that the Soviet space program was a natural expression of Marxist technocracy. I suspect it is too early to come to very definite conclusions about the Russian history of spaceflight, because reliable sources are only just becoming available, and we would expect some important parts of the picture to remain obscure for many years. McDougall's analysis of the early spaceflight societies is based entirely on secondary sources; and he is not very careful in using them. For example, he apparently slipped in cribbing from my book, because he got the date wrong for the founding of the American Interplanetary Society, 1926 instead of 1930, perhaps because he skimmed a page of my book rather than reading it closely (McDougall 1985, p. 26; Bainbridge 1976, p. 125).

A crucial empirical claim McDougall makes, apparently supporting his thesis, was that the Russian space club received significant government support and encouragement, even in its earliest days (McDougall 1985, p. 36). In part, McDougall is merely guessing, because primary sources on this phase of Russian space history are not yet available. And in part, he draws upon a clearly propagandistic book by Evgeny Riabchikov (1971) that offers the standard Soviet line of the late 1960s, with little in the way of factual support. Even Riabchikov (1971, p. 105) admits the Russian club "endured some severe hardships," and Michael Stoiko (1970, p. 47) says that "members were known to have been refused ration books because they were accused of being occupied with 'nonsensical fantasies.'"

Western intellectuals have often projected onto the Soviet Union their own fantasies of what a scientifically run society might be like. While not claiming the Soviet system was superior to the Western democracies and hinting that its technocracy was not necessarily benevolent, McDougall romanticizes Soviet technological history and did not seriously consider alternative interpretations. Had the Soviet Union been a real technocracy and really been evolving into a workers' paradise, I would have thought that substantial effort would have been invested into development of industrial production technology—robots and the like—but Japan and the United States led in this area. As I write, a remarkable year of events has torn apart the Soviet Empire, revealing endless surprises, and the reader will have the benefit of far deeper insight into its essential nature than I possess today, perhaps even gaining access to documents that establish beyond doubt how spaceflight developed behind the Iron Curtain.

Whatever the historical details of this period, a tremendous impetus to the Russian spaceflight movement came from the successful German development of the long-range liquid-fuel rocket. In 1945, the Russians seized the main V-2 development and production facilities, and in 1946 they kidnapped all the rocket engineers they could find in East Germany. Encircled by American airbases, rushing to develop atomic weapons to compete with the American bombs, the Soviet Union sought any means to outflank its enemy (Tokaev 1951). Without the example of the V-2, it would have had little reason to favor long-range rockets over other technologies.

Today, we all recognize the effectiveness of nuclear-tipped intercontinental missiles, but three facts about them should be kept in mind. First, the technically best form of war rockets is the solid-fuel variety, essentially an upgraded version of the gunpowder military rockets introduced in thirteenth-century China, but spaceflight requires the more powerful and controllable liquid-fuel variety that the social movement promoted. Second, a host of other means for delivering lethal warheads could have been developed as adequate substitutes for rockets; for example, the modern winged cruise missile is the descendent of the German buzz bomb, and experiments on robot aircraft were conducted as early as World War I. Third, large liquid-fuel boosters like Atlas and Titan, big enough to orbit manned capsules and send probes to the planets, were designed when fusion warheads were heavy, before technical developments reduced their weight until they could be hurled by small boosters of no use for space missions.

Thus, there was a crucial period in recent history—call it a launch window if you like astronautical metaphors—during which the application of liquid-fuel technology to long-range nuclear attack could benefit spaceflight. In taking a military detour, the spaceflight movement was racing against other technologies: the battlefield solid-fuel rocket, the cruise missile, and the compact fusion bomb. It won the race because effective individuals like von Braun were able to convince military planners to fund many expensive technical developments necessary for spaceflight and because world politics provided major international competitions that could be exploited.

Spaceflight technology is sufficiently advanced now to serve nonmilitary, utilitarian motives: communications, weather monitoring, and the like. The historical trajectory is like an orbital shot. Spaceflight took off in a tremendous blast of social energy. As it neared orbit, its thrust cut back considerably. Now it coasts, securely established in a parking orbit but lacking the power to take a new course to the planets. Some slow progress is probable, as near-Earth and synchronous orbits are exploited further, but a second great leap forward requires a second spaceflight revolution somehow ignited by the social movement that has always had its eyes on the stars.

This analysis suggests a conceptual distinction—not a logically perfect one, but useful—that can help us understand the results of my survey research. Following Kuhn, we can distinguish *normal* motives for spaceflight from *revolutionary* ones. Normal motives are utilitarian, either economic or military-political. They can be served by Earth-bound technologies; for example, fiber-optic cables present an ever-greater challenge to communication satellites. But because the first phase of the spaceflight revolution was successful, space systems are viable competitors with terrestrial systems.

Revolutionary motives demand a change in major aspects of

culture, society, or technology. They are the favorite goals of transcendent social movements, and they need not be related to space development. But if the spaceflight movement can connect many of them to grand new projects in the heavens, it could achieve another quantum leap. Often, revolutionary motives dissipate themselves in expressive frenzies, symbolic crusades, and supernatural revivals. But occasionally, like the Christian missionary spirit that assisted pure economic greed in opening parts of the New World, they can shape history.

In examining the values Americans find in the space program, we should distinguish the normal from the revolutionary goals. The former provide that basis of popular support required to maintain funding for a space program dedicated to serving conventional needs. And, therefore, the normal goals must receive high levels of acceptance to be effective targets of funding. In the past, revolutionary goals energized space development through a small, almost fanatical spaceflight social movement. Therefore, such motives need not become extremely popular to have a powerful effect, when amplified by unusual social conditions. To be sure, the more popular they are the wider their influence and the greater the chance they will play important roles in a future cultural transformation. But at the present, it is merely important that the more revolutionary goals are attractive to at least some people who might become influential, as individuals or groups.

Having established basic concepts, we can now begin our analysis of fresh data. Much of this book will focus on surveys I did at Harvard University, delineating the values of spaceflight as conceptualized by young members of the elite who have given the topic considerable thought. However, an essential preliminary step is to examine the views of average citizens, reflected in surveys limited in scope but administered to random samples.

STANDARD OPINION POLLS OF THE NATION

When national polls include a question or two about the space program, their aim is simply to document the level of support or weigh opinions on some current issue. Thus, their data offer few insights about the meaning spaceflight has for our civilization or the specific goals in space that ought to be achieved. However, a little can be learned from analysis of the levels of support offered by different groups in society, and the information to be gained is an essential basis for understanding the data from polls administered to specially selected sets of respondents.

In July 1944, the Gallup Poll included a pair of questions based on first reports about the German V-2 rocket program headed by Wernher von Braun: "A Swedish newspaperman says the Germans are now building robot bombs which can hit cities on our East Coast. Do you believe this is true?" "Do you think that in another twenty-five years such flying bombs will be able to cross the Atlantic Ocean?" Only 20 percent said yes to the first question, but 70 percent said yes to the second. The American public revealed good instincts in this poll, because in fact the Germans had only the most preliminary plans for rockets to fly farther than the 300-kilometer maximum range of the V-2, but twenty-five years later intercontinental rockets had long since achieved trans-Atlantic distances. The Gallup questions were not phrased in terms of rockets, and no connection with spaceflight was suggested, but the results indicate the wartime public saw aerospace technology developing rapidly (Gallup 1972, p. 456).

Less national prescience was revealed by a poll at the end of 1949. Although 63 percent felt trains and planes would be run by atomic power within fifty years and 88 percent predicted a cancer cure by the same time, only 15 percent felt, "men in rockets will be able to reach the moon within the next fifty years." The first moon landing came in only twenty years, rather than fifty, and at the forty-year mark nuclear trains and planes seem absurd and a general cancer cure remains remote (Gallup 1972, p. 875). By 1955, the percent feeling the moon could be reached in fifty years had risen to 38 (Gallup 1972, p. 1306). At that point, the first Earth satellite was only two years away.

Two weeks after Sputnik I, the Gallup Poll asked respondents in several cities around the world, "All things considered, do you think the earth satellite is more likely to be used for good purposes or for bad purposes?" Of the Americans polled in Washington and Chicago, 61 percent said good purposes, with 16 percent saying bad and 23 percent holding no opinion. In contrast, only 17 percent of respondents in Oslo, Norway, said good reasons, with 39 percent saying bad (Gallup 1972, p. 1519). Most of the post-Sputnik polls concerned the competition between the United States and Soviet Union, a topic we shall examine again in Chapter 6.

By July 1969, when men were actually landing on the moon, Gallup asked about the next goal: "There has been much discussion about attempting to land a man on the planet Mars. How would you feel about such an attempt—would you favor or oppose the United States setting aside money for such a project?" Thirty-nine percent favored the idea, with 53 percent opposed and 8 percent holding no opinion (Gallup 1972, p. 2209). These figures do not quite square with answers to a question in another poll that year: "The U.S. is now spending many billions of dollars on space research. Do you think we should increase these funds, keep them the same, or reduce these funds?" Only 14 percent wanted funding increased, whereas 40 percent wanted it reduced (Gallup Opion Index 1969). Apparently, many people wanted Mars without paying for it.

In 1973, a scientifically designed national poll, the General Social Survey, first asked a question on space funding, and this item has been included whenever the poll was given, usually every year (Davis and Smith 1986). Thus the GSS offers the opportunity to chart trends in the level of space support, and the many other items in the survey allow one to identify the segments of the public showing the greatest enthusiasm for spaceflight. The GSS space item is one of a set of eleven "problems" or "government programs," and the respondent is supposed to say whether too little, too much, or about the right amount is being spent on each: space exploration; improving and protecting the environment; improving and protecting the nation's health; solving the problems of the big cities; halting the rising crime rate; dealing with drug addiction; improving the nation's education system; improving conditions for Blacks;

the military, armaments, and defense; foreign aid, and welfare.

One of the most consistent findings of such polls is that men give far greater support than women. In 1985, immediately before the Challenger disaster, 15 percent of men felt too little was being spent on the space program, compared with just 7.4 percent of women. And 32.9 percent of men felt too much was being spent, versus 48.9 of women. That is, more than twice as great a proportion of men want funding increased, and women are more apt to want funding reduced.

The fact that there is a great gender difference in support for the space program tells us little about what spaceflight means to our culture. To label spaceflight *masculine* reveals nothing. We must ask which aspects of the differences between men and women—their upbringing, typical social roles, career expectations, modal personality constellations, value systems—are salient for spaceflight. Shortly we shall see that pro-space attitudes are held by the young and educated. Because they live longer, women respondents are slightly older than men, on average. Because equal educational opportunities are a recent phenomenon, fewer women have completed higher degrees.

Men may more often have technical careers, those that draw upon the same sciences and varieties of engineering that create spacecraft. Men more often enter the military and have positive attitudes toward it, a fact that is salient for the space program to the extent that people see it in military terms. Probably, all of these factors contribute, and others besides. I shall leave the job of conclusively explaining the gender differences through sophisticated statistical analysis for another time, because we have a different purpose here. When we examine the various aspects of spaceflight, as conceptualized by our culture, we shall occasionally look at gender differences. But the prime focus must always be on the reasons why people might support space, and the differential reactions to them of subgroups in the population are of secondary importance.

The gender difference holds for young people, as a 1980 national poll of teenagers showed (Gallup 1981). Sixty percent of boys felt the space program was a good investment, and 36 percent thought the money could be better spent on other problems. But a majority of the girls, 51 percent, favored switching the funds to other problems, and only 44 percent said the investment in space was good.

Despite the consistency of the gender differences across age groups, age is an important variable. The 1969 Gallup poll about an expedition to Mars found that 54 percent of respondents under age thirty were in favor, compared with 40 percent of those thirty to forty-nine and 28 percent of those over fifty.

To belabor the obvious, age is the result of two other variables: when people were born and when they were polled. And these two components of age are sociologically quite distinct. Do the young support space because they were born in more modern times than the old? If so, this is called a *cohort effect*, a characteristic of a group of people born about the same time (an "age cohort") because of the experience of growing up in their particular period. Or do the young support space because of youthful exuberance that will fade as they age? This is called a *maturational effect*, a difference between age groups caused by changes that affect each age cohort as it goes through the life cycle. We can put this in terms of predictions about the overall level of support shown by the entire society. As the years pass, will support for space grow as more and more people are born into the space age, replacing less enthusiastic people born before it? Or will support stay about the same, because an individual's enthusiasm fades with increasing age?

It is possible to explore this set of questions with the GSS data, because the surveys cover the span of a dozen years, a substantial fraction of a lifetime. We can look at how an age group's opinions change over time, to see whether a maturational effect makes older people give up former enthusiasm for novel projects like spaceflight. Table 1.1 does this, contrasting three age cohorts. People who were eighteen to twenty-nine in 1973 were polled at that time, and 10.2 percent of them wanted space funding increased. The same age cohort, all now a dozen years older, was polled again in 1985, and then 13.4 percent wanted funding increased. So, even though the group had aged, their level of support actually increased. (Note that the two polls did not survey exactly the same individuals, which would be the ideal procedure for research on changes over time, but the effect is almost the same.)

Table 1.1. Change in Support for the Space Program by Age Cohort

	Percent wanting funds for the space program increased	
Age in 1973	1973	1985
18–29	10.2	13.4
30–49	8.5	11.5
50–69	4.5	3.9

Similarly, the level of support for the space program increased for the group age thirty to forty-nine in 1973. Originally, 8.5 percent felt too little was being spent on space, and a dozen years later the proportion had risen to 11.5 percent. Clearly, neither of the two younger age groups lost enthusiasm for spaceflight, as would have been the case if a maturation effect were primarily responsible for the differences between age groups. The oldest group, those aged fifty to sixty-nine in 1973, shows a slight erosion of support, a drop from 4.5 percent to 3.9 percent. This could be the result of accidental fluctuations. Only 447 people were in this group in 1973, and the difference of 0.6 percentage points represents only 3 people. So, the old people hold steady at their low level of support, while the relatively high level of support of younger people actually increases.

Table 1.1 indicates the support given to spaceflight by the young is indeed a good omen, projecting a steady increase as new generations are born into the space age. The relevance for the meaning of spaceflight is simply that space is an aspect of modernity. As more and more people become true citizens of the modern world, socialized to the norms and values of advanced technical society, the support for spaceflight will grow.

National polls are not a perfect reflection of the strength of support spaceflight enjoys in the society, because decisions about space policy are not made by a random sample of the population. I shall not here enter into the acrimonious debate over the extent to which America is run by a "power elite" having only its own interests at heart, but clearly many segments of society have negligible influence, and those who have power are apt to have a different balance of views about the space program than those who do not. For example, voters are more positive than nonvoters. The 1985 General Social Survey asked respondents whether they had voted in the 1984 election, and an increase in space funding was approved by 11.7 percent of those who said they had, compared with 8.2 percent of those who had not, a ratio of 1.4 to 1.

Members of the upper social classes, those who presumably have more than their equal share of influence, tend to give more support to the space program than members of lower social classes. In 1985, 15.7 percent of those with incomes over $25,000 wanted space funding increased, compared with only 5.8 percent of those with incomes under $10,000. A very solid majority of the more affluent class wanted funding either increased or kept the same, 68.0 percent, whereas a majority of 54.2 percent of the poorer group wanted funding reduced.

We do not have data that would tell us conclusively which differences between the social classes are most responsible for the different attitudes toward the space program, but informed guesses are in order. The poorer groups may want government money spent on their own pressing needs, whereas the richer groups may feel that an economic surplus can be invested in future-oriented programs. The prosperous classes may identify more strongly with business and industry, appreciating the ways the space program can serve their interests. But more relevant for our study of conceptions of spaceflight, the upper social classes are better educated, on average, and thus both better informed about the space program and more fully committed to the intellectual gains it offers.

The social class difference begins early. The 1980 Gallup poll of teenagers found that parents' social class was a good predictor of young people's support for the space program. Although 58 percent of those from a white-collar background felt it was a good investment, only 48 percent of those from a blue-collar background agreed. Of those whose parents had attended college, 61 percent felt it was a good investment, compared with 48 percent of those whose parents had not. Interestingly, the students' own academic standing did not seem to matter, 53 percent of those above average and 52 percent of those average or below saying the investments in the space program were well spent.

In 1969, when Gallup sought people's opinions about a Mars expedition, 52 percent of respondents who had attended college were in favor, but only 39 percent of high school graduates, and 25 percent of those with less education. Table 1.2 shows levels of support by education, using the GSS data. In both 1973 and 1985, an absolute majority of those with little education wanted space funding reduced. Although the percent calling for an increase among those who had attended college increased only a little from 1973 to 1985, from 14.3 percent to 17.0 percent, the proportion wanting appropriations reduced shrank almost by half. Thus, for educated people, the value of the space program had been solidly established.

Table 1.2. Education and Support for the Space Program

Those feeling funds for the space exploration program are . . .	1973 national poll			1985 national poll		
	Some College	High School	Little Education	Some College	High School	Little Education
too little	14.3%	9.2%	2.5%	17.0%	12.1%	3.4%
about right	44.7%	36.2%	27.9%	60.5%	46.7%	38.6%
too much	41.0%	54.7%	69.6%	22.5%	41.2%	58.0%
Total	100%	100%	100%	100%	100%	100%
Respondents	217	719	552	324	792	412

We can compare these respondents with college students sampled in one of my own surveys. In 1981 I polled 1,465 University of Washington undergraduates. Although the sample was not strictly random, the research replicated results of surveys others had done with random samples, and the large number of items and respondents allowed me to determine that sampling bias was minimal. There was no General Social

Survey in 1981, but we can interpolate between the 1980 and 1982 polls. Support for the space program was at a high level in those years, 15.2 percent wanting appropriations increased. but the young college students were far more enthusiastic, 30.2 percent said that current funding was too little. Although 39.6 percent of the GSS respondents wanted space funding reduced, only 14.8 percent of the college students held this negative opinion. The greatest support came from the college men, 45.1 percent wanting funding increased, and only 8.6 percent, decreased.

A final question from the 1973 General Social Survey can help us understand the popular meaning of the space program, an item unfortunately not included in 1985. Respondents were supposed to say how much confidence they had in science: a great deal, only some, or hardly any. Whereas 11.6 percent of those with a great deal of confidence in science felt space appropriations were too little, only 2.1 percent of those with hardly any confidence in science felt this way. Indeed, 77.3 percent of these critics of science wanted funding reduced.

Much space science can be done without direct human participation. Indeed, a major policy issue of the 1970s and 1980s, which caused great ill feeling between various segments of the spaceflight movement, was the proper emphasis on manned spaceflight versus unmanned probes. The massive funding for the space shuttle came partly at the expense of planetary probes and other robot scientific missions. The public, of course, may not conceptualize the alternatives this way, and no election ever presented voters with the choice. A few national polls have explored the issue, although not in ways that revealed much.

Right after the Challenger disaster, *Newsweek* magazine commissioned the Gallup organization to poll 533 persons reached by telephone (Foley 1986). The most poorly phrased item was, "Do you think that putting civilians into space is important—or is it too dangerous?" Fifty-five percent said "important," 40 percent said "too dangerous," and 5 percent did not know. The meaning of the word *civilian* is quite ambiguous in this context, meaning either nonastronaut or nonmilitary, and important and dangerous are not logical opposites. Better stated was: "Some people say the United States should concentrate on unmanned missions like the Voyager probe. Others say it is important to maintain a manned space program, as well. Which comes closer to your view?" Although 21 percent wanted a completely unmanned program, 67 percent wanted manned as well. But respondents were not asked why they wanted a manned space program, and items like this are more tantalizing than satisfying, if we want to understand the popular ideology of spaceflight in any depth.

ADVANCED SURVEYS ON SPACE GOALS

To go beyond the limited findings of the national polls, research needs to employ more complex questionnaire items or extensive batteries of items measuring respondents' views on detailed aspects of the space program. Of necessity, this means abandoning expensive and uninformed random samples of the general population. For example, in 1963, Donald A. Strickland polled 211 physicists, asking them what they thought were the underlying motives of the American space program. They ranked international competition far ahead of other goals, 32 percent placing propaganda and prestige first, and 14 percent said that military motives predominated. Five percent each ranked exploration, basic research in natural sciences, or domestic political motives first, 4 percent placed economic motives first, and the remaining 35 percent wrote in another reply or failed to answer. Here the physicists were being asked to judge the goals societal leaders had chosen for the space program, rather than to set their own objectives, although their collective impression of political realities may not be far from the mark.

The most extensive early survey of space goals was a questionnaire administered by Raymond A. Bauer (1960) to 1,717 readers of the *Harvard Business Review*, most of them holding management positions. The response rate was about 31.5 percent, and like the Harvard student surveys I shall introduce shortly, pro-space people may have been more likely than others to respond. Therefore the data are better for exploring the early space goals of American business culture than for determining exactly the level of support given the space program by those in business. Eighty-five percent agreed that "outer space is the new frontier. Research and exploration will have profound and revolutionary effects on our economic growth." And 89 percent agreed that "mankind wants to go into outer space because it is there.... We are drawn by our desire to know and conquer anew." But only 9 percent agreed that a manned space program was unnecessary because robot machines could do the job required.

In response to a series of questions about the possible payoffs of the space program, 69 percent believed that revolutionary improvements in communications were almost certain to happen. Majorities were convinced that significant benefits were bound to come in the fields of medicine, biology, meteorology, robotics, mathematics, and physics. In contrast, only 4 percent felt that mining of other planets was almost certain to happen, and just 3 percent had the same confidence about colonizing other planets.

One set of items asked respondents to rank five possible objectives, reflecting the general reasons for supporting the space program Bauer was able to identify. Table 1.3 reveals that "pure science research and gaining of knowledge" was most often placed first, with a substantial number rating "control of outer space for military and political reasons" highest. Three years after Bauer's survey, Furash (1963) repeated the research with about 3,300 readers of the *Harvard Business Review*, and the same set of items was included in my 1986 survey of Harvard students, so I have included the distributions from these surveys as well. The Harvard students overwhelmingly rejected international competition as a goal in 1986, but gave about the same lukewarm response as did businessmen a generation earlier to the idealistic and emotional objective, "meeting the challenge and adventure of new horizons." In contrast, they rated scientific and economic payoffs much higher.

In 1977, I administered a survey to 225 registered voters who lived in Seattle, Washington, a study described in detail in the next chapter. One set of questions sought attitudes on the general goals of the space program, and they were introduced as follows: "Many reasons have been given for supporting the space program. Below are seven words describing different kinds of reasons. How good a justification is each one? Which are important reasons for continuing the space program? Please check the box after each one that indicates how important you feel it is."

Table 1.3. Rankings of Five Possible Objectives for the Space Program

	Percent rating the objective highest		
	Harvard Business Review		Harvard Students
	1960	1963	1986
Pure science research and gaining of knowledge	47	43	55.4
Control of outer space for military and political reasons	31	31	4.1
Tangible economic payoffs and research results for everyday life on Earth	14	18	30.5
Meeting the challenge and adventure of new horizons	8	8	9.2
Winning the prestige race with the Soviet Union	3	5	0.8

Majorities rated scientific and technological justifications as very important, 68.0 percent and 63.6 percent, respectively. Less than a third of respondents rated the others very important: economic (28.3 percent), social (15.7 percent), political (15.0 percent), psychological (12.6 percent), and religious (4.6 percent). Further, the scientific and technological values correlated highly ($r = 0.62$), indicating that respondents considered them to be very closely connected. Social reasons were connected about equally strongly with economic and psychological ones, achieving coefficients of 0.50 and 0.55. Thus, the data suggest that plausible justifications can be crudely divided into two groups, the very popular scientific-technological goals of spaceflight, and those with economic, social, and psychological implications.

The Seattle voter survey was but one of an entire series of such projects, spanning more than a dozen years, listed in Table 1.4. To identify each survey efficiently, in the pages that follow I use abbreviations like S1977. The *S* stands for "survey," and the numbers are the year it was administered. So S1977 is the space survey I did in 1977, which happens to be the study of 225 Seattle voters. When more than one was administered in a given year, I add *A* or *B*. S1986A and S1986B are the main surveys administered to Harvard students in the spring and fall of 1986.

Table 1.4. The Series of Space Goal Surveys

Survey	Respondents
S1973A	74 members of the New England Science Fiction Association
S1973B	80 participants in a Committee for the Future convention
S1974	102 members of the American Institute of Aeronautics and Astronautics
S1977	225 voters in Seattle, Washington
S1981	1,465 students at the University of Washington
S1983	212 Harvard University students
S1986A	1,007 Harvard University students
S1986B	894 Harvard University students

Social scientists have long recognized that little can be accomplished with the single questions about a public issue often incorporated in national polls. For example, sociologists of religion know that religiousness is a multidimensional phenomenon, and the survey researcher must use a complex battery of questions to measure its variations at all accurately (Glock and Stark 1965; Stark and Glock 1968). As Bauer (1969, p. 91) said about the success of his own questionnaire research on space attitudes, "the multidimensional approach is contrary to the established tradition of journalistic opinion polling, which has dominated our thinking on the sources of support for public programs. For reasons of economy, effort, ease of asking questions and of communication to the public, opinion pollers have regularly sought a single 'thermometer' type of question."

But if we abandon the journalistic commitment to single, simple questions and we cautiously relax the requirement of getting a random sample in the awareness that the average citizen does not have proportional influence over public policy, we can develop methods to gain much deeper insight about the conceptions of space held by American culture. And the full toolbox of survey techniques available to us must be used, qualitative as well as quantitative types of questions.

FUTURE SPACE PROJECTS

Single questionnaire items asking the respondent to check one of a small number of labeled boxes provide only limited insight into our culture's conceptions of spaceflight, even when the data are subjected to sophisticated statistical analysis. More difficult to analyze but exceedingly valuable for the ambitious researcher are nondirective questions that ask the respondent to contribute spontaneous thoughts. Items of the former type are usually called *fixed-choice, forced-choice,* or *closed-ended.* The latter type are called *open-ended.*

To illustrate the wealth of material than can be gained through open-ended items, I shall conclude this chapter by reporting responses to the following item in S1986A, my spring 1986 Harvard University survey: "Please briefly describe a future space project that you would like to see carried out." Altogether, 469 Harvard students wrote intelligible replies, and I have roughly categorized these responses in Table 1.5. Thirty-eight of the respondents gave two or more different projects, averaging 2.5 projects each. So that these exceptionally enthusiastic people would not unduly influence the distribution, their projects were counted fractionally. That is, if someone proposed two different projects, each was counted as ½ a project in the tabulation. Therefore, the percentages in Table 1.5 are based on votes cast by the 469 students, one vote per person, with each vote capable of being split between projects.

Table 1.5. Future Projects Proposed by Harvard University Students

Percent	Project
	Near Earth
16.7	Space station
5.1	Space telescope
4.6	Vehicle development
3.9	Medical and biological research
3.7	Solar power satellites
3.1	Food production
3.0	Civilians in space
2.5	Strategic Defense Initiative
1.5	Manufacturing
0.6	Earth research satellites
	Deep Space
19.4	Solar system exploration
9.2	Lunar and planetary colonies
8.7	Interstellar exploration
4.3	Orbiting cities
3.6	Moonbase
3.5	Extraterrestrial mining
2.5	Waste disposal
	Other
2.0	Communication with extraterrestrials
1.6	General science
0.5	Miscellaneous
100.0	Total (469 respondents)

I have divided the categories into three groups, those primarily describing activities in near-Earth orbit, those primarily concerning the Moon or more distant locations, and a few that cannot easily be categorized in terms of distance. This distinction is quite crude, but the two main groups are almost equal in size. Near-Earth projects were proposed by 44.7 percent of the 469 and describe work that can be done without crossing the orbit of the Moon.

Many Harvard University students believe it is time to build "a permanently manned space station," and the 16.7 percent in the space station category do not include the medical research station, solar power satellite, food production, strategic defense initiative, and manufacturing ideas, most of which assume the existence of such a station. According to explanations offered by respondents, among its functions would be maintenance of "satellites used for weather and T.V. communications," "learning more about the rest of space unimpeded by the Earth's atmosphere," "zero-gravity testing," and "developing prototypes for space-based industry." Various opinions held it should be "open for habitation by civilians," "totally self-supporting," and "a cooperative Soviet-American space station." One Harvard student urged "space stations which can replicate themselves and thus extend into space, after many years."

Enthusiasm ran high for an orbiting telescope, and several respondents were clearly aware that one was currently awaiting a launch vehicle, two identifying it correctly as the Hubble Space Telescope. Others called for instruments to follow it: "bigger space telescope (after we get the first one up)," "very large multiple mirror telescope," "radio telescope arrays," and "further deployment of space telescopes—both optical and in other regions of the spectrum." One proposed "a group of large space-based telescopes, spaced some distance apart. Interferometry allows you to combine them into one effective instrument with large resolving power."

In the vehicle development category, four people simply wanted "more space shuttles." Related ideas were "air-space-planes" for fast "ground-to-ground travel" on Earth, "trans-atmospheric vehicle, horizontal take-off," and "spaceships that could land on regular airstrips." A relatively modest project, the "geostationary platform," would involve a structure in synchronous orbit to take the place of many separate communications satellites. More ambitious, would be "a viable shuttle to the Moon." Really radical vehicles mentioned were "rocket boosters with miniatomic explosions" and "the 'space elevator' program like Arthur C. Clarke describes in *Fountains of Paradise*."

"I'd like to see a bio lab set up in space for the advancement of medicine," the "establishment of permanent medical research center in space—manned or unmanned." "A medical treatment facility on the Moon" might improve "medical treatment efficiency, e.g. slow down of heart rate, hasten healing in different atmosphere." Basic research could be done on such questions as "birth and growth of animals in space," "geotropism/geotaxis in plants/animals in the absence of gravity," or "man's physical and mental capabilities in space." A "medical laboratory in space" could engage in "manufacture of medicines in microgravity which can't be produced on Earth."

"Solar energy, collected in space stations, transferred to Earth" was frequently endorsed. The system would involve "solar collection panels in space collecting energy" "to be beamed to substations on Earth."

The idea of "food production" in "farms in space" or "in space stations" was not well developed, and undoubtedly it represents an attempt by respondents relatively unfamiliar with the space program to enlist it in the service of pressing human needs. "Space agriculture" might produce "more nutritious foods which can not be grown on Earth" to "help countries such as Ethiopia" and alleviate "the problem of world hunger."

The proposal to expand civilian travel into space took many forms. Four students wanted to go into space themselves, and another said, "I'd like to see a group of schoolchildren take a field trip in space." Other possibilities included, "civilian travel to see the Earth from space," "the construction of a resort complex in space," "sightseeing shuttles to the Moon for affordable rates," and "family trips to the Moon." These students seemed to feel that space should not be reserved for the astronauts, but opened "for all American citizens."

Eleven students urged military projects, five of them simply writing "SDI" (Strategic Defense Initiative). Others wrote "ABM [antiballistic missile] defense," "Star Wars," "military satellite," "anti-nuclear defense station in space," and "defense against a nuclear attack." A more grand proposal urged "establishment of U.S. military bases on the Moon or in space with the capability of launching vehicles which could engage and destroy."

"Space factories producing products" "in a zero-gravity environment" were mentioned by five students. They could undertake "crystal growth in space, self-supporting" economically and partly doing the "manufacturing in space to avoid pollution of Earth." The small Earth research category proposed "more satellites collecting oceanographic data" and "study of the Earth's atmosphere as a radiation shield—how it preserves life here."

Deep space proposals, those that imply travel beyond low Earth orbit, constitute 51.2 percent of the total. Substantial numbers support a manned base on the Moon, exploration of the solar system, establishment of colonies on the Moon or planets such as Mars or in free-flying space cities like those proposed by the L-5 Society. The far-off goal of exploration beyond the limits of the solar system also receives support. Clearly, a portion of the Harvard student body wants a greater role for human beings beyond the immediate vicinity of Earth.

The large solar system exploration category might be divided into three kinds of proposal: those for unmanned probes (11.3 percent) , those for manned expeditions (6.1 percent), and a remainder that cannot be classified. Many agreed with the student who said, "I think the Voyager program was great," and they wanted "further Voyager probes." Frequently mentioned targets were the sun, Venus, Jupiter, Saturn, Titan, Pluto, comets, and the Oort Cloud (hypothesized to be the home of the comets). It is time to launch "a suicide probe into the sun which would send back data as long as possible, until it burnt up." "Send a probe to Venus" for "thorough robotic exploration" and "photographs/analysis of the surface." We could learn much from another "probe exploring in turn Jupiter, the Galilean moons, Saturn and Titan, and Neptune and Triton in much greater detail" than achieved by Voyager II. Even more exciting would be an "unmanned probe to land and retrieve samples from Io, Titan, other moons of Jupiter and Saturn." We might also consider "a well organized search for microbial life in the Jovian clouds/moons."

The manned exploration subcategory would have been larger, except that lunar proposals usually went into the Moon base category, and many Martian proposals urged outright colonization. Ten people wanted "a manned Mars landing," perhaps in the form of "U.S.-Soviet exploration." But the ideas in this category make up in grandeur anything they lack in number. Respondents foresaw a majestic "planetary 'grand tour,' as with Voyager but manned." This would be a "multiyear manned mission" fulfilling "a joint Soviet/American role, taking Soviet/American children with them" to "the moons of Jupiter, Saturn and Uranus." This might be done with "large, volunteer-manned self-sustaining space stations that will travel and explore the solar system." One student said: "I would like to see a space station *not* in Earth orbit. Perhaps a highly elliptic orbit that brings it through the orbits of several planets as well as the asteroid belt."

"Thorough exploration" of the solar system, whether by humans or machines, should involve "studying planets up close and directly to give us more clues to life and the evolution of our solar system." "There should be "exploration of planet surfaces and compositions of atmospheres," "analysis of plate tectonics on other planets," and "more exploration of planets that can sustain life."

"Colonization" or "settlement of the Moon" seemed a feasible goal, perhaps "by 2100." "It is a small gravity well, a mineralogically wealthy place, and a great takeoff place." "I would like to see industry on the Moon taking advantage of an uninhabited, resource-laden satellite, instead of draining the Earth dry of her resources." With "colonization of the Moon and planets," "overpopulation might become an obsolete problem!"

Perhaps not fully aware of the difficulties involved or willing to consider very long-range projects, respondents proposed "interstellar travel" and "exploration of other solar systems." Some suggested unmanned probes: a "satellite to Alpha Centauri," one "to go to Barnard's Star," or even "a probe that would examine the nature of black holes." Others urged "sending manned ships to other solar systems." For a "manned expedition to Alpha Centauri, multi-generation" ships would be required, "perhaps some sort of extremely large (hollowed-out asteroid) manned craft." One thought "fusion drive" would be sufficient, but five felt a real "starship" would require a new principle of propulsion, "faster than light drive so we can expand out into the galaxy." More practically, one urged "the BIS Daedalus project," a proposal by the British Interplanetary Society to build unmanned interstellar probes.

Six students wanted a space colony built at the L-5 point in the Moon's orbit, and two would be happy with a different Lagrange point, a concept pioneered by physicist Gerard K. O'Neill (1977) and publicized by the L-5 Society. It should be "permanent, independent" largely "self-sufficient," and perhaps "built using mass driver on Moon," the materials being shot from the lunar surface by a magnetic catapult. One felt it was important "to develop liveable stations in space so that man can escape a possible terrestrial disaster." Another wrote, "I wish to see, within my lifetime, an orbiting space colony —like the one depicted in Disney's Epcot Center." If we combine this orbiting city category with the one about lunar and planetary colonization, then colonization of space was the favorite project of 13.5 percent.

One might also want to include the lunar base category, bringing the colonization total to 17.1 percent, but the phrases *Moon base* and *station on the Moon* do not necessarily imply extensive habitation and economic exploitation, and proposals vary in the magnitude of the project suggested. "We need a space lab on the Moon for astronomical observations" or a "permanent Moon base for space/scientific/commercial research." "A permanent station on the Moon. Such a station would be of great scientific value as well as a convenient launching point for future exploration." "Long-term Moon base with hundreds of people on it. Self-supporting in all ways."

Some people thought "asteroid mining" or "mining of the Moon" would be feasible. Either we could establish "stations in the asteroid belt for mining the asteroids," or start "towing asteroids back to Earth for precious metals." We should also "search for more natural resources on other planets."

If problems of launch safety and cost could be solved, "storage of nuclear and chemical waste in space" might be possible, "perhaps relocating toxic waste to the lifeless zone of some asteroid" or "using the Moon as a cosmic space dump." "I'd like to see some means developed to get rid of the wastes we are currently dumping into the seas." Four students suggested "nuclear waste disposal in the sun," as the sun is a vast nuclear furnace that would not be affected by the radiation. One even suggested "a probe sent into the sun, and one day carrying the Earth's nuclear weapons with it."

The remaining 4.1 percent of the proposals could not be categorized either as near Earth or deep space. In principle, communication with extraterrestrials could occur in deep space or on the Earth. Indeed, as we shall discuss in Chapter 8, a serious search for radio signals from extraterrestrials has for several years been carried out by professor Paul Horowitz of Harvard University, and one S1986A respondent actually mentioned his name. Others urged "continuing attempts at radio contact with life on other planets," "unrelenting effort to contact other inhabitants of the galaxy," and "peaceful contact with alien civilization."

The general science category included "survey of high energy particle emitters in space," "study of sunspots and coronal activity," and "tests of 'physical laws' under other conditions." One student hoped that space research could determine whether the current expansion of the universe might reverse. "The most important question I have that hasn't been answered is whether there is enough matter in the universe to pull it back together again." The most alarming of the miscellaneous proposals said, "The possibility of using nuclear explosives to alter the trajectory of an object which presents a threat to the Earth might be tested out on an asteroid."

A CHALLENGE FOR SOCIAL SCIENTISTS

Clearly, articulate members of our culture possess more complex thoughts and opinions about space than can be captured in one or two check-the-box questions. And, as evidenced by its paucity of publications, the sociology of space is still in its infancy (Cheston, Chafer, and Chafer 1984). To chart fully the conceptions of spaceflight held by late twentieth century Americans, we need extensive survey research combining both qualitative and quantitative data. How we can accomplish this is the theme of the next chapter which describes my pilot research and explains an effective survey methodology.

A grand vision of rapid space progress is painted by the

report of the National Commission on Space, published just months after the Challenger disaster, ironically within a few days of when Challenger had been scheduled to hurl the long-delayed Galileo probe to Jupiter. In its summation, the commission listed five general goals: scientific knowledge, technology advances, national leadership, new opportunities (for personal fulfillment, enterprise, and human settlement), and "hopes and dreams inspired by removing terrestrial limits to human aspiration" (National Commission on Space 1986, p. 192). In the following pages we will learn whether these goals are reflected in the values our culture has imagined for spaceflight.

CHAPTER TWO

THE PILOT STUDY

All great human enterprises are primarily social, and the space program is no exception. It could not exist without social support, without the enthusiasm of national leaders and the acceptance of the general public. Great technical advances would be impossible if no one were willing to pay for them. To use the metaphor of war: The attempt to conquer space will be won or lost on the home front. World War II was decided as much in the factories as on the battlefields; the Vietnam War was a disaster of the spirit, not a failure of technology. This chapter places both technology and human spirit under the lens of ethnographic survey research. With data from the spaceflight movement and the general public, it analyzes a coherent set of justifications for the space program—the Ideology of Spaceflight.

Too many nonsociologists, and sadly too many professionals in the field, believe it is easy to create a successful questionnaire. One sits down, they imagine, in a mood of creativity and begins to write. Questions flow from the fingertips into the word processor, and in a few days—behold, a high-quality questionnaire. In fact, the best professional researchers spend as much time and effort creating their questionnaires as analyzing them. They draw questions from earlier research, when appropriate, often investing months in a careful review of the literature of their field. If new items are to be included, then questions painstakingly composed to examine the topic under scrutiny are subjected to a lengthy process of pretesting, both as separate items and as batteries that make up measurement scales. Although for decades questionnaires have occasionally included items about spaceflight, some cited in the previous chapter, we really do not yet have a flourishing tradition of survey research on the subject, and thus I found it necessary to undertake a multistage research program.

In the mid-1970s, I administered a series of surveys in two main phases that constitute my pilot project. The twin surveys done at Harvard University in 1986 provided the main dataset on which this book is based, and I shall consider them in detail beginning with the next chapter. But the pilot study was so rich in its findings that it deserves close attention as well. Indeed, it worked so well that if I did not have a larger dataset to consider, I would not designate it a *pilot*, but present it as a valid, complete piece of research.

In mid-1972, I attended the world science fiction (SF) convention held in Los Angeles, and interviewed fifty-eight participants to learn their views of the value of the space program. In 1973, I mailed a short questionnaire to members of the New England Science Fiction Association (NESFA), receiving replies from seventy-four of them. One question asked, "In your opinion, what is the most important reason why we should continue the space program?" Respondents were given five lines on which to write their answers.

Although there is good reason to blame science fiction for giving the public a distorted image of our opportunities in space and even for creating negativism when we fail to accomplish all that has been predicted in its wild stories, still it has done much to popularize spaceflight and recruit thousands of persons to its cause. In 1986, Harvard University Press published my *Dimensions of Science Fiction*, a study of the ideologies of science fiction based on survey data, including much about its relationship to spaceflight. The fantasies of science fiction writers are no substitute for the sober work of aerospace professionals, but they do play a role in shaping public conceptions and setting long-range goals for the space program.

For part of my dissertation, I examined a colorful space-boosting organization, The Committee for the Future. Founded by Barbara Marx Hubbard, this uninhibited group was planning a grand future for the human species, including colonization of the planets, and at one time it actually attempted to buy a leftover Saturn V rocket for its own moon mission. I may have been a bit harsh on the CFF in my book, as Michael Michaud has observed:

> To William S. Bainbridge, the history of the CFF suggested that there was little or no opportunity for amateurs to participate in furthering the exploration, exploitation, and colonization of the solar system. Yet the CFF was a harbinger in many ways, particularly in its efforts to project a broad, positive vision of the future that fully incorporated the spaceflight revolution, and to involve ordinary citizens in the pro-space cause. (Michaud 1986, p. 45)

In my analysis, the CFF provided an extreme contrast to the private spaceflight groups of the 1920s and 1930s, such as the German Society for Space Travel and The American Interplanetary Society, which were able to achieve major advances in rocket technology despite their extremely limited financial and social means. By the early 1970s, I found it difficult to see what practical engineering projects a private group could carry through successfully, and this, in part, was what the CFF seemed to want to do. A related question, ably examined by Michaud in his book but still beyond my own research program, is the influence private space-boosting groups may have on public opinion or on decisions by policy makers.

The Committee for the Future represented the radical wing of the movement, and its rhetoric was thus worth including in my research on the justifications for spaceflight. I culled tape recordings I had made at two CFF conventions, in 1972 and 1973, for any statements participants had made about the goals of the space program. I did the same with a stack of the group's literature I had collected. And eighty participants at the 1973 convention in Washington, D.C., filled out a questionnaire I mailed to them, including the open-ended question about the most important reason for continuing the space program.

If SF and the CFF could provide imaginative and radical goals for the space program, the right place to look for more conventional and realistic goals was at the heart of the aerospace industry. I chose to examine views of the American

Institute of Aeronautics and Astronautics, the most prestigious large scientific-engineering organization in the aerospace field. From a random sample, 102 members responded to a questionnaire in 1974.

In addition to the item from the SF and CFF questionnaires, the AIAA survey contained other open-ended items designed to collect thoughts on the possible value of space: "Suggest another reason for space progress that is a good selling point for convincing the intelligent layman to support the space program." "Some perfectly valid and important justifications for the space program are often ignored and deserve greater mention than they commonly receive. Can you give us such a justification?" "Can you name a bad consequence which would follow if progress in space exploration and development were halted at the present level?" "What is a possible long range result of a vigorous space program, which would eventually be significant for mankind?"

The work with the AIAA, the CFF, and the SF subculture produced a tremendous amount of data. To analyze it, I went through all the recordings and questionnaires, copying down every statement that was an answer to the implicit question, Why should we continue the space program? Each statement was typed on a file card, a total of 1,256. The largest number, 620, came from members of the AIAA, whereas 340 were contributed by science fiction fans and 296 came from the Committee for the Future. I refer to these statements as *utterances*, using the word as a technical term. Each one is a distinct idea from an individual person, but they are not all complete sentences. A person might say, "Well, I think communication satellites are a good thing, and I believe weather satellites have great benefits, too." These are two utterances, one about communication satellites, and another about weather satellites, and they belong on separate cards.

The next step was to reduce the 1,256 utterances to a more manageable number. Many of the cards really said the same things. For example, several mentioned something about communication satellites, and another group cited weather satellites. I carefully sorted the cards into piles that seemed to be expressing single main ideas, winding up with forty-nine. In work such as this, one has to find a balance between lumping and splitting. *Lumping* is placing many utterances in the same category, even if they are only slightly similar in concept. *Splitting* is dividing the utterances into separate categories on the basis of even very slight differences in meaning. Lumping produces far fewer categories than splitting. For the pilot study I got 49 categories, and for the main study described in Chapter 3, admittedly based on a substantially greater number of utterances, I got 125.

Using wording on the cards, I wrote a summary statement on behalf of each group. For example, one pile concerned weather satellites and generated the following simple summary: "Meteorology satellites aid in making accurate predictions of the weather." But this could be considerably expanded, using statements on cards in the pile. "The satellite has proved an effective tool in increasing the reliability of short-range weather forecasts," and "meteorological satellites have saved hundreds of lives by hurricane warnings that were formerly not possible." "More than forty nations benefit from the improved weather forecasting based on cloud-cover photos relayed by a space system equipped with satellites." "If you can predict better the weather on the whole Earth, you could save something like $2 billion annually." "Studying of seasonal changes can lead to prediction of spring floods through knowledge of the northern snow cover." "More sophisticated satellites may someday help predict and track tidal waves, earthquakes, rainfall, floods," and "perhaps even lead to reduction of human life losses due to tornadoes."

THE 1977 VOTER SURVEY

A set of forty-nine paragraphs like this could serve as the manifesto of the spaceflight social movement. I call such an essay, created by the social scientist through surveys and interviews, a *synthetic ideology*. Despite shelves of enthusiastic space-boosting books, a full ideological statement of the movement's beliefs has never been made. This book is a step in that direction, but I did not stop with the summary statements and their piles of amplifying utterances. Instead, I built a new questionnaire around the forty-nine summary statements, which I call S1977, because it is the space survey I administered in 1977.

After a brief section of general questions, the survey introduced the space goals with this introduction. "On the following pages we have listed forty-nine reasons that have been given for continuing the space program. You will probably feel that some reasons are better than others. Don't worry about all the aspects of each one, but make an overall judgment of it. After each statement please check the box that indicates your opinion. How good a reason is it for supporting the space program?" Four responses were offered: not a good reason, slightly good reason, moderately good reason, and extremely good reason. There were five versions of the questionnaire, identical except that the space goals were in five different random orders.

With the great assistance of Richard Wyckoff, I mailed this survey to a random sample of registered voters in Seattle, Washington, receiving analyzable replies from 225 voters, representing a response rate of about 45 percent. We did not use a follow-up procedure to coax more persons in the random sample to respond, and I doubt that most people have the knowledge to give high-quality responses to questions such as these. The main role the respondents were to play was that of judge, giving careful consideration to the forty-nine space goals and through their responses helping us discover the connections between them. As I shall explain further in Chapter 3, for such research a perfect random sample is not crucial and usually impossible to obtain.

The Seattle voters held a wide range of opinions, but there were consistent patterns. Some of the statements got a very favorable response, whereas others received poor ratings. Communication satellites headed the list, with 68.8 percent of the voters calling them an "extremely good reason" for continuing the space program. Lowest on the list with only 4.5 percent, was the statement: "Without spaceflight we would be trapped, closed-in, jailed on this planet." Appendix A of this book lists the forty-nine space goals, along with the relative popularity of each and other statistics, and again I must thank Richard Wyckoff for his tremendous help with the quantitative analysis at a time when I had not yet mastered computers.

But the questionnaire was designed to be more than a popularity contest. I primarily wanted to discover the general concepts behind all the many specific ideas about the value of space. That is, I wanted to delineate the conceptual structure of the spaceflight ideology. How do different ideas cluster together? What principle unites all the ideas in a cluster? What underlying values are served by the conquest of space? These questions can be answered by several statistical methods designed to process vast amounts of information and identify dimensions and groupings. This chapter will use one called *factor analysis*, and the following chapters will employ one known as *block modeling* with a far larger dataset.

Factor analysis begins with the opinions of our 225 voters on each of the forty-nine space goals, a total of roughly 11,000 pieces of information. It first produces a matrix of correlations between all pairs of items, expressing the degree of association linking each pair. For example, the correlation between the two extraterrestrial life items is 0.59, and that between communication with extraterrestrials and military applications is dead zero. The basic correlation matrix contained 1,176 such numbers.

Then the computer was instructed to look for factors in this matrix. Each factor can be understood in two different ways. It could be seen as a conceptual dimension along which all forty-nine space goals are arranged, each having its own location on the dimension. Or, perhaps more profitably for present purposes, it can be seen as a cluster of intercorrelated items that elicit similar patterns of response from the voters. The computation procedure is entirely mechanical, but I inspected several such analyses, cutting the structure into few or many pieces, comparing the solutions to see which one did the best job of clustering ideas into mathematically sound and conceptually meaningful groups.

The solution we shall consider most closely called for five dimensions, instructing the computer to produce five varimax rotated factors, and five very clear groups emerged. For example, four space goals having to do with international and military competition were placed together. Weather satellites and communication satellites went together in another factor. We shall discuss the five in order of descending popularity, beginning with one that showed an exceedingly high level of support from the voters.

THE INFORMATION FACTOR

Eleven space goals grouped together, most sharing a common element of information gained or transmitted through satellites, are listed in Table 2.1. The first column of figures shows the *factor loadings*, the statistics that let us place these items in the same group. When the computer produced an analysis with five factors, it printed out a table containing five columns of loadings, one for each dimension. With only slight simplification, we could consider the loadings to be the correlations linking each item with the cluster as a whole or with the underlying but unmeasured concept that links all the items brought together into the cluster. Items with high loadings are more closely associated with the factor. Items with low loadings are not members of it.

In two brief articles that reported a few of my findings (Bainbridge 1978a; Bainbridge and Wyckoff 1979), a space goal was counted as a valid member of a factor only if its loading was 0.45 or more. This is a bit arbitrary, and the reader might prefer to set another criterion, so here I will proceed slightly differently. Each of the forty-nine space goals will be assigned to the factor on which it achieved its highest loading. Forty pass the 0.45 criterion, and the lowest loadings counted are 0.36. In reading Table 2.1 or any of the four that follow, emphasize in your mind the goals with higher loadings and consider dropping from consideration those at the very bottom. As I contemplate the lists, it seems to me that even the lowest loaded items share something with the others in their group, although they may stray far from the central idea.

Table 2.1. The Information Factor

	Factor Loading	Popu-larity
5. Navigation satellites are a great help to ship and plane navigators, and traffic control from space can aid safe and efficient use of conventional transportation systems.	0.68	76.8%
6. Meteorology satellites aid in making accurate predictions of the weather	0.67	81.5%
4. Earth resource satellites allow us to monitor the natural environment of the Earth and help locate valuable resources such as minerals and water.	0.65	83.3%
13. Space technology produces many valuable inventions and discoveries which have unexpected applications in industry or everyday life.	0.62	81.1%
14. Space development will give us new practical knowledge that can be used to improve human life.	0.60	75.9%
3. Radio, telephone, and TV relay satellites are vital links in the world's communications system, fostering education and international understanding.	0.60	88.9%
15. Space technology will allow us to manage the environment of our planet because it is developing techniques for managing artificial environments that support human life.	0.55	57.4%
21. Space exploration adds tremendously to our scientific knowledge.	0.52	85.7%
48. Space will be of value in ways we cannot yet imagine.	0.43	74.1%
7. Electric power generated in space and sent down to Earth will help solve the energy crisis without polluting our environment.	0.39	64.1%
37. Space can provide a focus for increasing international cooperation leading to world unity.	0.36	65.0%

The goals in the Information factor were exceedingly popular with the Seattle voter respondents. The indicator of popularity given in Table 2.1 is the percent judging a goal to be either "extremely good" or "moderately good." By this measure, only seven of the space goals achieved popularities above 75 percent, one-seventh of the total forty-nine, and all are in this group. The average popularity of the eleven Information goals is 75.8 percent. None of the four other factors had a popularity above 50 percent.

Four of the six most strongly loaded Information goals identify types of Earth satellite that long ago achieved practical benefits: navigation satellites, meteorology satellites, Earth resource satellites, and communication satellites. The other two speak more generally about knowledge, discoveries and inventions: "Space technology produces many valuable inventions and discoveries that have unexpected applications in industry or everyday life." "Space development will give us new practical knowledge that can be used to improve human life." The loadings for these six are bunched tightly together, from 0.60 to 0.68, and they equally define the factor, all refer to the gathering and transmission of information.

Two other goals come close behind in the loadings, techniques for managing the environment at 0.55 and scientific knowledge at 0.52. The first refers to a specific area of know-how developed in the space program and transferred to more mundane uses, and thus it is an example of the spin-offs mentioned in general terms by two of the goals above it. The second is the popular item citing the general benefit of increased scientific knowledge.

Three other items with low loadings but high popularities complete the factor. The proposition that "space will be of value in ways we cannot yet imagine" expresses a faith that harmonizes well with the high popularity of the factor. It is gratifying to see that nearly 75 percent of respondents feel this is a moderately or extremely good reason for continuing the space program. The item about electric power generated in space suggests a future way that orbiting satellites may benefit citizens of Earth. The final item, about international cooperation, has such a low loading that we can hardly call it a member of the factor. Indeed, it achieves loadings above 0.25 on three other factors.

THE ECONOMIC-INDUSTRIAL FACTOR

Six space goals belong to the second-most popular factor, receiving a 48.0 percent rating, and are listed in Table 2.2. The first pair, with substantially higher loadings than the rest, speaks of economy and technology. Three others focus on space-related careers, and the fourth resists losing the capabilities and know-how possessed by space professionals. Loading equally at 0.49, the goals in this quartet correlate most highly with the top pair of items. I suppose we could call this the Technology-Career factor, although the name I first proposed when the research was done is also appropriate, and thus there is no need to replace it.

Table 2.2. The Economic-Industrial Factor

	Factor Loading	Popularity
17. The space program provides an essential stimulus to the whole economy by investing money and paying employees.	0.63	40.0%
16. We must continue the space program in order to maintain the quality of American technology.	0.59	50.2%
26. The space program must be continued so that its highly trained technical manpower will not be wasted in unemployment.	0.49	26.3%
28. The space program encourages young people to choose careers in science and technology and is itself a good training ground for scientists and engineers.	0.49	59.8%
24. The space program must be continued so we do not lose the capabilities we have developed for spaceflight and the rocket know-how we have invested so much to gain.	0.49	51.6%
29. Space progress will provide new, rewarding jobs for many people.	0.49	59.8%

Had they not already been included in the Information factor, the following items would also have been included here: "Space technology produces many valuable inventions and discoveries that have unexpected applications in industry or everyday life." "Space exploration adds tremendously to our scientific knowledge." The spin-off item loads 0.62 on Information and 0.45 on the Economic-Industrial factor. Scientific knowledge gets 0.52 and 0.45. Thus, both are not far behind the quartet at the bottom of Table 2.2. Even though we do not list them in the Economic-Industrial factor, they influence the loadings of the other items, and thus the group has even a bit more of a technological flavor than the table demonstrates.

THE MILITARY FACTOR

Four items about military applications and international competition, joined by another sharing a common element of hazard with them, constitute the Military factor shown in Table 2.3. The average popularity of the five is 43.4 percent, and the two most highly loaded define the group well: military applications and international competition. The first four items are favored by political conservatives and disfavored by political liberals, the only ones in the set of forty-nine that show a substantial correlation with a question on political orientation I had included. Therefore, the popularity of this cluster is highly dependent upon the mix of political persuasions among the particular set of respondents polled.

Table 2.3. The Military Factor

	Factor Loading	Popularity
1. Space has military applications; our nation must develop space weapons for its own defense.	0.77	44.2%
27. Space is an important arena for international competition, and if we do not keep our lead, the Russians will gain an advantage over us.	0.74	40.6%
25. The success of the U.S. space program increases our prestige in the world, demonstrates the value of democracy, and renews American national pride.	0.61	42.9%
2. Military reconnaissance satellites (spy satellites) further the cause of peace by making secret preparations for war and sneak attacks almost impossible.	0.56	59.1%
10. Outer space will be used in the disposal of very dangerous waste products, such as unwanted radioactive materials.	0.41	30.1%

The item about extraterrestrial disposal of radioactive materials and other dangerous waste products actually correlates most highly (0.41) with an item in the Colonization factor, which recommends doing dangerous experiments in space. Further, its loading on the Colonization factor is 0.33, compared with 0.41 on Military. Its average correlation with the Colonization items is 0.26, compared to 0.29 with the other Military items. Thus, it could have been placed in Colonization almost as easily as here, but it would not be entirely comfortable in either factor.

THE COLONIZATION FACTOR

Fourteen space goals listed in Table 2.4, with an average popularity of 33.3 percent, constitute the Colonization factor. The first two, tied with loadings of 0.70, explicitly talk about colonizing other worlds, whereas the third, close behind them, describes the societies we might make there. Four identify activities humans may conduct in colonies: exploiting raw materials, operating hospitals, commercial manufacturing, and performing dangerous experiments. Most remaining items express instead the basic values that might motivate or legitimate colonization: escaping the confines of a terrestrial trap, ensuring survival of the species, expanding for societal health, continuing economic growth without limit, and communication with extraterrestrials.

Table 2.4. The Colonization Factor

	Factor Loading	Popu- larity
20. Overpopulation on Earth can be solved by using the living space on other planets.	0.70	24.9%
32. Space travel will lead to the planting of human colonies on new worlds in space.	0.70	24.3%
34. Society has a chance for a completely fresh start in space; new social forms and exciting new styles of life can be created on other worlds.	0.66	24.0%
8. Raw materials from the moon and other planets can supplement the dwindling natural resources of the Earth.	0.63	50.9%

Table 2.4. *(continued)*

	Factor Loading	Popu-larity
33. Our world has become too small for human civilization and for the human mind; we need the wide open spaces of the stars and planets to get away from the confines of our shrinking world.	0.59	17.6%
38. Spaceflight is necessary to ensure the survival of the human race against destruction by natural or man-made disaster.	0.57	25.6%
30. Human societies have always needed to expand in order to remain healthy; space is the only direction left for such expansion.	0.56	31.4%
19. We must go beyond the finite Earth into infinite space in order to continue economic growth without limit.	0.54	20.7%
11. Space hospitals put into orbit where there is no gravity will be able to provide new kinds of medical treatment and give many patients easier recoveries.	0.53	50.7%
9. Commercial manufacturing can be done in space without polluting the Earth; completely new materials and products can be made in space.	0.47	40.6%
35. Communication with intelligent beings from other planets would give us completely new perceptions of humanity, new art, philosophy, and science.	0.44	55.3%
22. We can conduct certain dangerous kinds of scientific experiment far in space so accidents and other hazards will not harm anyone.	0.42	36.2%
46. Without spaceflight we would be trapped, closed-in, jailed on this planet.	0.41	14.7%
12. Rockets developed for spaceflight will be used for very rapid transportation of people, military equipment, or commercial goods over long distances on the Earth.	0.38	49.1%

Although the lowest-loaded item refers to rocket transport back on Earth, it expresses enthusiasm for a kind of travel essential in space colonies. Its loading on the Information factor is practically identical with that on the Colonization factor, 0.36 compared with 0.38, and it describes a mode of communication between points on Earth—direct travel. Respondents were probably unaware of the early experiments with rocket-delivered mail, but they undoubtedly see a connection between terrestrial rocket transport and satellite communication of pure information.

Three colonization goals have popularities above 50 percent, and two others are above 40 percent. Each of these five speaks of benefit for Earthlings from activities in space, whereas the other nine colonization goals are focused on the colonies themselves. Even the top item, about solving overpopulation on Earth, directs attention away from our globe toward other planets. The average popularity of the five terrestrial benefit colonization goals is 49.3 percent, whereas the nine more extraterrestrial goals that remain achieve only 24.4.

THE EMOTIONAL-IDEALISTIC FACTOR

The final group in the five-factor analysis, close behind Colonization in average popularity, is a set of thirteen emotional or idealistic goals, given in Table 2.5. They speak of personal feelings like curiosity, excitement, loneliness, love, and the desire to fly into outer space. But there are also values on the societal or species level, including global renewal, a goal for mankind, a substitute for war, and human destiny. Several of these items can be read as referring to either the individual or the group, promising challenge and expansion of mind and spirit for person and species alike.

Neil Armstrong's famous first words from the lunar surface were revealing: "One small step for a man, one giant leap for mankind." They remind us both that space achievements are accomplished by individuals only as members of a vast team, and that the glory belongs to all of us, not just to those who actually fly the missions. The space goals of the Emotional-Idealistic factor link positive experiences individuals gain from spaceflight to positive moral lessons learned by all humanity. Through them, person and society are transcended.

Table 2.5. The Emotional-Idealistic Factor

	Factor Loading	Popularity
42. We must explore space to satisfy our great curiosity; space exploration is an expression of man's natural curiosity.	0.74	46.8%
40. Space exploration is an exciting adventure, valuable for the fun and excitement it provides.	0.69	20.3%
36. Human society on Earth needs the change and global renewal that space travel will bring.	0.65	19.1%
31. Men and societies have always needed the challenges provided by a frontier; space is the new frontier.	0.64	52.4%
41. We must explore space for the same reason people climb Mount Everest—because it's there.	0.60	25.1%
45. Spaceflight enlarges the mind and the spirit of man, so that his ideas become universal rather than Earth-bound.	0.57	45.6%
44. Space can provide us with a goal and sense of purpose which mankind badly needs.	0.55	26.9%
47. Mankind needs to know it is not alone in the universe, both to gain humility and to lose the feeling of loneliness.	0.52	18.6%
39. I am in favor of the space program because I would very much like the experience of traveling into space myself.	0.52	22.5%
43. Travel into space will teach us to love and respect our own planet, Earth.	0.50	40.6%
23. Space exploration must continue so we can learn if there is life on other planets.	0.47	43.7%
18. Space exploration is a positive substitute for war because it channels man's aggressive instincts into non-military activities.	0.46	44.0%
49. In entering the universe mankind fulfills its destiny—to mature as a species and perhaps come closer to God.	0.36	17.1%

Two of the Colonization goals achieved loadings above 0.40 on the Emotional-Idealistic factor, 0.43 and 0.42, respectively: "Society has a chance for a completely fresh start in space; new social forms and exciting new styles of life can be created on other worlds." "Communication with intelligent beings from other planets would give us completely new perceptions of humanity, new art, philosophy, and science." People who favor the goals in the Emotional-Idealistic factor seem especially open to contact with strange forms of life, whether biologically alien or culturally radical.

OTHER FACTOR ANALYSES

The brand of analysis that I used with the S1977 data permits one to specify, within certain limits, the number of clusters into which the computer will divide the items. We have been considering an analysis with five factors, a moderate number, but there is nothing sacred about five. It is wise to experiment cutting the data into different numbers of pieces, to see how robust are the main factors and to hunt for lesser structures of meaning that may have been overlooked. Consequently, we will now examine analyses calling for four, six, and ten factors, for comparison with the five just seen.

Moving to a smaller number of factors typically squashes together some groups with fewer constituent items, but it may also bring together groups with a strong connection that producing more factors earlier may have fractured. Going from five to four factors leaves the Information, Colonization, and Emotional-Idealistic factors unchanged, exactly the same items being loaded 0.45 or better.

The four top military items remain together, but they steal two items from the former Economic-Industrial factor: "We must continue the space program in order to maintain the quality of American technology." "The space program must be continued so we do not lose the capabilities we have developed for spaceflight and the rocket know-how we have invested so much to gain." It seems to me that these harmonize with the international competition and unremitting struggle of the Military factor.

Thus, the chief result of reducing the number of clusters from five to four is the loss or fragmentation of the Economic-Industrial factor. The four space goals from this group that did not move in with the military items, do not achieve high loadings anywhere. One way of understanding this is to say that they represented a true, independent cluster, which could not be squeezed in elsewhere when dispossessed from its rightful spot.

In a six-factor analysis, the Military and Emotional-Idealistic factors remained unchanged from the five-factor solution, and only slight shifts occurred in the Information and Economic-Industrial factors. Colonization changed more noticeably, losing items to the extra factor, which consists of two items: "Commercial manufacturing can be done in space without polluting the Earth; completely new materials and products can be made in space." "We can conduct certain dangerous kinds of scientific experiment far in space so accidents and other hazards will not harm anyone." Their average popularity is 38.4 percent. The common element here may be preservation of the terrestrial environment, although this hypothesis is contradicted by the insignificant loading of 0.17 achieved by the item about space technology allowing us to manage the environment of our planet. Perhaps the Environmentalist Movement of the 1970s had both made protection of the ecology salient for respondents and stated the issue in terms of leaving Earth alone rather than managing it. We should look for this Environmentalist factor in later analyses.

The final analysis of the S1977 data expands to ten factors. This should give every solid conceptual fragment a chance to emerge. Two of these ten were essentially empty, no space goal achieving a loading even as high as 0.35. This indicates that with eight factors we have reached the limit in the number of groupings the computer could find. Most items keep their familiar places in our five main factors, but three other small groups appear: two pairs of goals and a triplet.

One factor is the Environmental pair discussed earlier. The other pair combines the following goals, which had originally been associated with the Information factor: "Space exploration adds tremendously to our scientific knowledge." "Space can provide a focus for increasing international cooperation leading to world unity." Their average popularity is 75.4 percent, essentially the same as the 75.8 of the original Information factor. On the surface, it is not clear what holds these two together. The goal of scientific knowledge has high correlations with many of the others, achieving coefficients above 0.45 with thirteen goals from three of the original factors: Information, Economic-Industrial, and Colonization. Thus, this goal falls between these factors when given the room to do so. The international cooperation item does not achieve as many high correlations, its strongest being 0.47 with its partner in this small factor, scientific knowledge. It correlates above 0.40 with eight other goals from four of the original five factors, excluding the military one.

Thus, one thing that brings scientific knowledge and international cooperation together is the fact that each of them is pulled toward several different large factors. But they must also have something in common, because of the significant correlation linking them, and international cooperation is more closely connected to scientific knowledge than to anything else. One possibility is that Seattle voters were responding to the internationalism of science itself. Not only are scientific research and communications carried on in a more international manner than almost anything else one can mention, but scientists have been in the forefront of promoting better international relations. Indeed, the internationalist Peace Movement has had prominent scientists at its head for many decades, from the physicist Albert Einstein and the mathematician Bertrand Russell of decades ago to the many scientists active in antinuclear movements at present. But, again, one should not make too much of this idea in the light of the breadth of connections that link these two goals to those in other factors.

Finally, we have a clear factor of three Extraterrestrial Life items: "Communication with intelligent beings from other planets would give us completely new perceptions of humanity, new art, philosophy, and science." "Space exploration must continue so we can learn if there is life on other planets." "Mankind needs to know it is not alone in the universe, both to gain humility and to lose the feeling of loneliness." Their average popularity is 39.2 percent. Although NASA has sometimes provided modest funding for communication with extraterrestrial intelligence (CETI), also holding symposiums and issuing publications on the topic, it is not a central thrust of the space program, and there is no thought that mere rocket flight will achieve CETI. However, both as a long-term goal and a focus of public interest, it deserves analysis, which we shall provide in Chapter 8.

THE UNIVERSITY OF WASHINGTON STUDY

In 1981, I had the opportunity to expand on the 1977 voter survey. With sociologist Robert Crutchfield, I received a small grant to do a criminological survey (Bainbridge and Crutchfield 1983). For me, the real purpose of the project was to piggyback some spaceflight items in a varied questionnaire administered to a large number of respondents. The questionnaire was completed by 1,465 undergraduates at the University of Washington in Seattle, 55.3 percent of them women.

The University of Washington is the dominant educational institution in the American Northwest, training ground of the future elite of an entire region of the nation. Although my 1,465 respondents are not a true random sample of the student body, having been recruited primarily through introductory sociology classes, questions designed to measure nonrepresentativeness indicate there are no serious biases (Bainbridge and Crutchfield 1983). If anything, persons likely to concentrate in the social sciences are underrepresented, as these sociology classes were among the most popular ways an undergraduate could fulfill the social science requirement for graduation. Not only has good previous research been based on the same population and used the same recruitment method, but a large 1979 study of 1,439 Washington undergraduates examining ideological orientation in depth found the same pattern as for a random sample of youth in the San Francisco area (Bainbridge and Stark 1981a).

The questionnaire was primarily about matters other than space, and therefore the respondents are not self-selected space boosters. Thus, it is interesting that they were rather supportive of the space program, a point made in the previous chapter, in their responses to the GSS space funding item. One item sought respondents' opinions on the following statement: "In the long run, discoveries made in our space program will have a big payoff for the average person." The majority, 52.7 percent, agreed. Only 21.6 percent disagreed, and the rest were neutral.

Two other agree-disagree items were phrased negatively. "The United States is spending too much money on space, so appropriations for the space program should be reduced." "Space exploration should be delayed until we have solved more of our problems here on Earth." Only 17.6 percent called for a reduction in appropriations, and 60.6 percent explicitly disagreed with this proposition. A somewhat larger group, 22.6 percent, agreed that we should delay space exploration, but a majority of 54.1 percent rejected delay. Even if one were to place those who responded "neutral" in the negative category, majorities support the space program.

Twenty statements adapted from S1977, presented as agree-disagree items, are listed in Table 2.6, along with the percentage who agreed with each. The items are grouped according to the results of a factor analysis I ran on them. I must stress that factor analysis is a tricky procedure which can fail for a variety of reasons. In particular, it is very important to begin with an unbiased set of items that accurately and evenly represent the range of concepts that respondents have. Never is this ideal achieved perfectly, but the 1977 voter study came close because its forty-nine items were based on a careful ethnographic study of ideas about spaceflight available to members of our culture. But, in reducing the number to twenty, I was unable to preserve exactly the balance and the scope of the original set. Indeed, I selected them partly on the basis of my then current research interests, rather than as an intentional replication. Thus, we cannot expect a perfect duplication of the earlier results, but we may get close.

The factor analysis was based on the 1,344 students at the University of Washington who expressed opinions on all twenty of the space goals. Because of the limited number of items, I did not demand that the computer find five factors, but let it extract the number with eigenvalues greater than 1, a conventional criterion. The result was four varimax rotated factors. As

Table 2.6. Space Goals in the University of Washington Study

	Factor Loading	Percent Agree
INFORMATION:		
Radio, telephone, and TV relay satellites are vital links in the world's communication system.	0.57	92.8
The space program can help maintain and improve the overall quality of our technology.	0.55	72.4
Meteorology satellites aid in making accurate predictions of the weather.	0.54	75.4
Space technology produces many valuable inventions and discoveries which have unexpected applications for industry or everyday life.	0.54	77.6
Space exploration adds tremendously to our scientific knowledge.	0.53	84.7
Earth resource satellites allow us to monitor the natural environment of the Earth and help locate valuable resources such as minerals and water.	0.47	69.8
Space will be of value in ways we cannot yet imagine.	0.47	80.0
IDEALISTIC:		
If we abandon the space program, we will be giving up on the whole idea of human progress.	0.59	56.7
Space exploration must continue so we can learn if there is life on other planets.	0.57	43.6
We have always needed the challenges provided by a frontier, and space is the new frontier.	0.55	65.6
Spaceflight enlarges the mind and the spirit of mankind, so that our ideas become universal rather than earth-bound.	0.52	58.1
Space exploration is a natural expression of innate human curiosity.	0.36	81.3
The space program provides an essential stimulus to the whole economy by investing money and paying employees.	0.33	48.2
COLONIZATION:		
Space travel will lead to the planting of human colonies on new worlds in space.	0.67	50.2
Society has a chance for a completely fresh start in space; new social forms and exciting new styles of life can be created on other worlds.	0.59	47.1
Raw materials from the moon and other planets can		

supplement the dwindling natural resources of the Earth.	0.53	47.9
Electric power generated in space and sent down to Earth will help solve the energy crisis.	0.45	46.9
Communication with intelligent beings from other planets would give us completely new perceptions of humanity, new art, philosophy, and science.	0.42	61.6

MILITARY:		
Space has military applications, and we must develop space weapons for our own defense.	0.56	33.4
Military reconnaissance satellites (spy satellites) further the cause of peace by making secret preparation for war and sneak attacks almost impossible.	0.54	47.5

was the case when we looked at four factors in S1977, the Economic-Industrial factor is the one missing. One item from it wound up in the Idealistic factor, presaging a result we will find in the 1986 Harvard University research: "The space program provides an essential stimulus to the whole economy by investing money and paying employees." But this item is at the bottom of its factor in terms of loading and got nearly the same loadings from two other factors. Another Economic-Industrial item originally read, "We must continue the space program in order to maintain the quality of American technology." Feeling that this was phrased in a too imperative and too nationalistic manner, I revised it greatly for the 1981 survey: "The space program can help maintain and improve the overall quality of our technology." In this form, it wound up near the top of the new Information factor.

Most popular of the factors, as before, was Information. On average, 79.0 percent of the 1,465 respondents agreed with each of the seven statements that constitute it. The top five items are bunched closely together in their loadings, indicating that they equally represent the key idea. The addition of technology, it seems to me, does not greatly alter the meaning, and thus I have kept the name Information for the factor. But, as always, the reader is encouraged to do his or her own thinking about what these items have in common. At the bottom, they draw in

"Space will be of value in ways we cannot yet imagine." This item has a loading of 0.40 on the Idealistic factor, and 0.37 on Colonization, suggesting it is not exclusively a member of this group. Fully 80.0 percent of the respondents agreed with this general expression of faith in the space program.

I call the second-most popular factor *Idealistic*. Before, we referred to an Emotional-Idealistic factor, but with the exception of an item about curiosity, the very emotional justifications for spaceflight are absent from our list, and thus could not shape the meaning of the group. On average, 58.9 percent of respondents agree with each of these six items.

Third in popularity is the Colonization factor. On average, 50.7 percent of respondents agree with each of its five items. Note that this is a majority, and that substantial proportions of those not agreeing with each colonization item said they were "neutral." Again, the respondents were not selected because they had a special interest in space, and thus these high levels of support for so radical a goal as colonization of the planets are impressive. As before, the item about finding life on other planets went into Idealism, whereas the one about extraterrestrial civilizations went into Colonization. Biology and civilization are two different things. Electric power generated in space wound up in Colonization here, although it had a tenuous hold on Information in the earlier study. Linking Earth with its space colonies, this space goal is appropriately tied to both, as we shall see in the Harvard University results.

Finally, the two military items composed the fourth factor. The statement about military applications gets a split response, almost equal thirds of the students agreeing, neutral, or disagreeing (33.4 percent, 34.9 percent, 31.7 percent). Reconnaissance satellites, with their explicit goal of ensuring peace, get a more favorable response, only 26.6 percent disagreeing.

Supplemented by S1981, the S1977 dataset identifies a clear structure in spaceflight ideology. Depending on the degree of detail we demand in our analysis, we find some or all of the following factors: Information, Economic-Industrial, Military, Colonization, Emotional-Idealistic, Extraterrestrial Life, and Environmental. Space goals that promise practical, near-term benefits to inhabitants of Earth are most popular, but respondents give substantial support even to such distant goals as colonization of other planets and communication with extraterrestrial beings. Both Seattle voters and undergraduates in the chief university of their area agree strongly that "Space will be of value in ways we cannot yet imagine."

THE QUALITY OF QUANTITATIVE DATA

The research process can be brought full circle. We can take the final quantitative results and return with them to understand better the ideas that people expressed in the qualitative, open-ended surveys. For example, we can ask again whether the three space-related groups I initially surveyed gave significantly different justifications for spaceflight.

The five groups of items in our main analysis of S1977 can be divided conceptually into two supergroups. The Information, Economic-Industrial, and Military factors represent relatively *normal* justifications. The Emotional-Idealistic and Colonization factors are more *revolutionary* justifications. In my book on the social movement that produced modern rocketry, I pointed out that small, radical groups had developed spaceflight for revolutionary reasons, to transport society to the planets and achieve a vastly different kind of existence there (Bainbridge 1976). But today's spaceflight establishment, represented by the AIAA, is oriented toward the conventional aims of spaceflight, using space to influence earth-bound society in modest ways. Thus, we would predict that the AIAA stresses conventional justifications, whereas the science fiction respondents and the CFF stress revolutionary justifications.

To test this eminently reasonable hypothesis, I returned to the cards from which the space goals in the 1977 questionnaire had been drawn. Admittedly, I used somewhat different methods to collect the utterances from the three groups, so statistical comparisons are not entirely trustworthy. But the group of respondents most dedicated to modest, *normal* technological development in space, the AIAA, had been challenged by questions demanding a spectrum of responses—including possible long-range benefits. And in the context of a pilot study the differences between the groups seemed amply documented.

A total of 1,042 utterances had contributed to the forty goals that achieved loadings of 0.45 or greater on one of the five factors in our main analysis, and here I will restrict myself to them. For example, consider the statement, "Space technology produces many valuable inventions and discoveries that have unexpected applications in industry or everyday life." There were 113 cards in the pile that gave birth to this item in the questionnaire, most highly loaded on the Information factor, ninety of them from AIAA respondents, eight from CFF respondents and fifteen from SF respondents. This conventional idea came primarily from the most conventional of the three space-oriented groups.

Altogether, 409 utterances had gone into the goals in the two revolutionary factors, 192 into the Colonization factor and 217 into the Emotional-Idealistic factor. And these came disproportionally from the CFF and the SF subculture. Only 20.2 percent of the 539 AIAA utterances had gone into them. But 57.3 percent of the 278 CFF utterances and 61.5 percent of the SF utterances did so. Thus, the majority of CFF and SF ideas that wound up in the factors can be described as revolutionary.

The space goals favored by the radical SF and CFF respondents were rated low by Seattle voters and University of Washington students. At the same time, the conventional goals of the aerospace establishment, represented by the AIAA, received high ratings. Although the public may not be willing to pay for a great thrust forward in space development at the present time, it does want gradual expansion of our spaceflight capabilities. And a few of the more extreme goals, such as communication with extraterrestrials, fascinate substantial numbers of ordinary Americans.

The data show quite clearly that the wings of the spaceflight movement are functioning well to identify and publicize goals for the space program. The conservative core of the spaceflight culture, the AIAA, provided us with a wealth of space goals that can be described as practical benefits. The radical SF and

CFF did their proper job as well, expanding the list to include a number of much more revolutionary goals. The Spaceflight Movement needs both the practical engineer and the radical dreamer performing complementary functions. The fact that space goals clustered into logical, cohesive factors, in the S1977 and S1981 datasets, proves that many in the educated general public understand the basic motives for spaceflight and are able to express coherent opinions about them.

Were this not the case, there would be little reason for continuing in the present line of research. But the pilot study has confirmed the scientific value of examining public conceptions of spaceflight in some depth. Our methods of research work well. With confidence, we can explore much larger datasets from the main surveys. As we do, we shall consider an even greater body of pro-space rhetoric, perhaps of use to any members of the movement who read this book. From the scientific standpoint, we shall be alert to any differences we find and evaluate whether the main study's findings replicate those from the pilot study. And in the combination of all the surveys and analyses, we shall learn the concepts and structure of American spaceflight culture.

CHAPTER THREE

THE MAIN SURVEYS

After completing the research described in the previous chapter, I launched a survey study of the concepts and values held by members of the science fiction subculture, represented by those attending a world science fiction convention held in Phoenix, Arizona (Bainbridge 1986). I was interested in the competing views of the future of science and technology expressed in this popular form of literature, and I was convinced that documenting them would contribute indirectly to an understanding of the cultural basis of space exploration. But much of my work in the late 1970s and early 1980s was devoted to the sociology of religion, culminating in a pair of books written in collaboration with Rodney Stark (Stark and Bainbridge 1985; 1987). Then, in 1986, a combination of opportunity and accident returned me to the study of popular conceptions of spaceflight.

In January 1986, I joined the press corps that gathered at Jet Propulsion Laboratory for the Voyager II close encounter with the planet Uranus. Having observed the 1981 flyby of Saturn, I had developed a loose plan to document the way the scientists and scientifically sophisticated journalists cooperated to interpret the meaning of the Voyager mission.

In his remarks welcoming the press, Dr. Lew Allen stated space goals.

We are very conscious of the fact that these flights are paid for by the taxpayer. They have, of course, the purpose of gaining in scientific knowledge, of contributing to prestige in the United States' exploration of space. But the purpose in part is certainly to allow the public to share in the excitement of the voyage: to learn of new worlds, to lift the spirits of people, to encourage them in inquiry, to inspire their search for further understanding.

On January 28, many of us sat at our desks in the pressroom, watching NASA's direct television feed from Cape Canaveral on the monitors that had been showing us the latest pictures from Uranus. It was the launch of Challenger. The three broadcast networks had chosen not to show the event live, considering the flight of a space shuttle too routine to let it disturb the daily schedule of soap operas and quiz shows. Then came the explosion.

The people at JPL understood immediately the significance of what had happened. Some personally knew one or more of the seven astronauts killed in the explosion, and all understood the devastating blow that had struck a space program already crippled by national disinterest and a budget too small to accomplish its mission. The Voyager spacecraft that was the center of JPL's attention had been launched nine years earlier, and JPL had not sent a probe to the planets for eight years. The bright hope of the planetary exploration program was Galileo, a delay-plagued project to place a robot spacecraft in orbit around the planet Jupiter, scheduled for launch by the space shuttle in May 1986. Everyone at JPL knew the explosion would delay Galileo at least two years, and over the following months their fears that the delay would lengthen to infinity came close to realization.

In the hours that followed, each person at JPL found his or her own response to the Challenger tragedy. In public, each did the necessary tasks related to Voyager with admirable skill and efficiency. Only in private did their grief and frustration burst forth. My response, the modest sociological contribution I could make, is this project to clarify public conceptions of our goals in space. By the end of the day I was designing my next survey.

The design was based on the work described in the previous chapter. I would do a pair of questionnaires. The first would collect a large number of ideas about the possible value of spaceflight. The second would get respondents' reactions to these ideas to provide the quantitative material for an analysis of how the ideas fit together and how they rest on more general concepts. One possibility would be to follow the pilot study exactly, obtaining the ideas from aerospace professionals and members of the Spaceflight Movement, then getting the quantitative data from a random sample of the general public. But I had already done that in the earlier research, and other possibilities existed.

Instead, I could do the entire research project with NASA employees or with a combination of spaceflight professionals and members of space-boosting organizations. They could provide not only the ideas in a qualitative survey but the statistical material in a quantitative questionnaire as well. I still think this would be a fine project. But I was not convinced that 1986 was a good year to do it in. Although the whole nation knew the space program's survival was in doubt, among aerospace professionals and spaceflight enthusiasts, the Challenger disaster was so devastating that I doubted I would get stable judgments. There were such agonizing debates, such deep depression, and such desperate searching for a way to rebuild the space program that people were just not in good enough condition to respond calmly and reliably to my questions. At least, such were my worries about a project that depended upon these respondents.

A project along these lines, however, would have a major scientific advantage over the pilot study. Both phases of the survey would use the same pool of respondents. If we want to study the ideas that exist in a culture or subculture, it is best to design the quantitative part of our study entirely using responses drawn from that same group. Thus, to take ideas from aerospace people and get reactions to them from the general public is interesting, but not methodologically pure. If one wants to learn how the general public might respond to new ideas, then this approach is fine. But if one is documenting the conceptualizations of a culture, it is better to stick within its boundaries.

To do a pair of surveys with random samples of the American public would be prohibitively expensive, indeed an awful waste of money. National polls using long questionnaires cost hundreds of thousands of dollars, and the research would require two of them. Although I am happy to have data from occasional national polls that include one or two space questions, I have severe doubts about the value of a national poll containing a hundred or more questions on space. Most people simply think little about space, and they know even less. The pilot study had achieved a response rate of 45 percent, pretty good without a follow-up, and had produced sound results. But Seattle is an aerospace town, and the survey had gone exclusively to registered voters, people especially likely to have and express opinions. More important, the pilot study had not expected the voters to provide the forty-nine space goals in the survey, merely to react to them.

I decided to focus on a specially selected set of respondents, and I was extremely lucky to have one readily available to me. At the time, I was enjoying a five-year visiting appointment at Harvard University. I had already done a small survey obtaining students' views on extraterrestrial intelligence, data that will figure in Chapter 8, and it was clear that Harvard students were an ideal population for research.

Harvard is among the top universities in the world, by any measure, and its students go on to influential careers in business, politics, science, and intellectual affairs. Highly meritocratic, Harvard represents a major segment of the American elite—indeed the elite of the world. The average voter is seldom given an opportunity to shape space policy; it is hard to mention an election that has depended upon candidates' plans for the space program. But as future policy makers and opinion leaders, Harvard students will contribute much to the decisions that will take us upward to the stars or leave us without an effective presence beyond the Earth.

Furthermore, these students have thoughts about spaceflight and a range of opinions on it they are ready to express. The purpose of the survey would not be to obtain a random sample, anyway, but to get the ideas and reactions of people who wanted to express them. Respondents in the first phase of the research would be like a think tank, suggesting possible goals for the space program. Respondents in the second phase would be a panel of judges, evaluating the proposed goals and identifying connections between them. Neither respondent role requires a random sample. Both require keen minds, a wealth of ideas, some factual familiarity with the issues, and the willingness to devote an hour of mental labor to the questionnaire.

Logistically, the initial step after creation of the first questionnaire, which I call S1986A, was to obtain the permission of the individuals and committees responsible for ensuring the safety of human subjects and of students at Harvard. The human subjects committee for the social sciences was quick to give formal permission, and I also obtained approval from two separate deans' offices. Approval also came from the masters of the Harvard houses. Essentially all undergraduates live on campus, including many whose families are in the Boston area. Freshmen room in Harvard Yard, and the upperclassmen live in a dozen residential complexes, called *houses*, set up almost like independent colleges. For both phases of the survey, the masters of the houses and house offices gave not only their permission but vital assistance in distributing the surveys.

My method of reaching the students was simple but extremely effective. Over the course of the project, I or one of my students visited every dining hall repeatedly, taking over a table immediately adjacent to the desk where each student must check in. Advertising ourselves with posters and signs designed carefully to attract attention without prejudicing students for or against any particular aspect of the space program, we offered questionnaires and pencils. We did this at every house, at the freshman dining hall, and at a dining hall that serves the few students who live in town. Most students who filled out the survey did so over an extended dinner, but a few took the surveys away with them to complete. Students interested but too busy to do the survey the first time we appeared always had a second chance later.

In each phase of the research, we continued until we had obtained a thousand surveys. Each undergraduate had the opportunity to get a survey, and no major category of undergraduate had a noticeably greater or lesser chance to get one. In this, the survey was like a random sample. But most students who accepted the challenge of my rather long questionnaires were people with definite thoughts about the space program that they wanted to express. These were people willing and able to volunteer as brainstormers and judges, to help me discover the goals of spaceflight, as conceived in the American intellectual culture of which Harvard is so central a part.

S1986A, administered in March, April, and May, got responses from 1,007 students. Actually about 10 percent did their job only halfheartedly, checking boxes dutifully when the question demanded that, but writing in very little. Thus the research relied most heavily on the roughly 900 who wrote extensive ideas. S1986B, the quantitative questionnaire based on the ideas in S1986A, was administered in October, November, and December. Again, about 10 percent of the respondents did not fully accept the challenge, leaving some questions unanswered. Because my tools of statistical analysis work best when few data are missing, I set a criterion level for rejecting questionnaires that were not sufficiently complete. Any S1986B questionnaire with more than 5 answers missing out of the total 147 was dropped, and my statistics are all based on the 894 respondents who answered essentially all the questions.

To be perfectly clear on how the research was carried out, we must first examine S1986A and see how the goals of spaceflight were derived from it.

DISCOVERING SPACEFLIGHT GOALS

S1986A contained two sections primarily designed to elicit utterances about the value of the space program and various aspects of it. One contained eight fixed-choice items, with the instructions: "First, check the ONE box that comes closest to your own, personal opinion. Then, write a brief comment explaining your answer in the space below the question, giving us a bit of your thinking on the particular issue." These issues included manned versus unmanned spaceflight, the space station project, the Strategic Defense Initiative, the exploration of Mars, and the far-out idea of communication with extraterrestrial intelligence.

The second section of S1986A consisted of open-ended items, with the following introduction:

Ideas About the Space Program.
  The following eight questions ask you to write brief statements expressing your thoughts. Unlike the questions above, these seem to assume you
have a favorable attitude toward the space program—but please do not be
put off by this. The purpose of this section is to collect many ideas about
why people might support the space program or various space projects.
All these ideas will be sifted carefully, compared with each other, and
incorporated in a future questionnaire that will assess the enthusiasm (or
lack of enthusiasm) that Harvard students have toward each of them.
Thus, we would greatly appreciate your responding to these questions,
whether or not you personally support the space program.

The first four of these items gave us the material for defining
space goals. "In your opinion, what is the most important reason why we should continue the space program?" "Can you
mention a very different benefit of the space program?" "Some
perfectly valid and important justifications for the space program are often ignored and deserve greater mention than they
commonly receive. Can you give us such a justification?" "Can
you mention a possible long range result of a vigorous space
program that would eventually be significant for humanity?"
  When I had obtained responses from a thousand students, I
went through all this material identifying distinct ideas, what
in the previous chapter I called *utterances,* which were printed
out on paper and cut into separate slips, one for each. Following the method developed in the pilot study, I then sorted these
utterances, repeatedly, into piles, arriving at a total of 125, each
containing utterances from at least two respondents. As the correlations from S1986B will show, I probably produced more categories than necessary, but the large number would permit a
more sensitive analysis. As earlier, I wrote a summary statement expressing the main idea in each pile of utterances, and
these became the 125 items about space goals that constituted
most of S1986B.
  To illustrate how the process worked, I will summarize the
material that went into item number 33: "Technological spin-offs
(advancements developed for the space program, then applied to

other fields) improve everyday life." This space goal was derived from 115 utterances that came from 107 respondents. Thirty respondents actually used the term *spin-off*, often coupled with *technological*, while several offered synonyms like *side benefits, off-shoots, by-products, fallout, boons,* and *technological trickle down.*

Spin-offs were described as "incidental developments," "external benefits," and "tangential" to the main purpose of the space program. Several respondents provided a definition. "Spinoffs—technology developed by the space program which can be used in everyday life." For one respondent they were "advances in technology when applied to other fields," and for another, "results that overlap into other areas."

Other respondents mentioned specific examples, including nonstick frying pans, super technological household items, mirrored eyeglasses, disposable diapers, microchips, miniaturization, calculators, computers, batteries, better television technology, new production techniques, laser technology, developments in medicine, energy, alloys, stronger but lighter metals, and "space-age materials." Six people mentioned Tang, the orange drink consumed by the Apollo astronauts, and four nominated velcro. One said, "Mylar plastic was first used for spacesuit helmet visors—now it's an everyday transparent and unbreakable plastic." A few of the examples were stated humorously. "Where would eggs be today without Teflon?" "Dried ice cream is great. Keep up the culinary breakthroughs."

The respondents may not be correct in citing specific spin-offs. For example, although velcro was widely used in the space program and its application to consumer goods was probably stimulated by publicity thus generated, it was not actually invented for astronauts. A Swiss engineer came up with the idea in 1948. Similarly, we cannot expect all of the space goals developed in our surveys to be technically feasible. Rather, the questionnaire responses express the thoughts of many Americans, and our respondents are representatives of American culture, not aerospace experts.

Some briefly analyzed the role of spin-offs in modern society. "Most of today's high-tech hardware—from microwave ovens to moonbuggies—is the direct result of space and military research." "Advances in space technology lead to advances in civilian technology." "Because large amounts of money are used for research, many new products develop for practical use." "New technologies resulting from the space program are often put to a myriad of practical uses in society." "Like all research and exploration, there are unforeseen side benefits in useful everyday items." "The space program provides a good catalyst for science." "It is a testing ground for technology to use here." "It stimulates, more than almost any project, development here on Earth for peaceful purposes." In similar manner, each of the other 124 goals emerged from reading a set of statements, and I shall quote the most interesting of them in appropriate places throughout the rest of this book.

Survey S1986B began with the comment, "Today, there is much public discussion about the future course of the American space program. This questionnaire seeks your thoughts on this important topic, and while your participation in our research will be completely anonymous, your ideas and attitudes will contribute to the current debates." It went on to explain that the questionnaire was based on open-ended answers from 1,007 students the previous spring. "Over the summer, we analyzed those written responses in great detail. One main result was a list of 125 somewhat different justifications for the space program, expressed in the actual words of Harvard students, and these became the heart of the present survey. Now we need you to rate each of these ideas, so we can compare their levels of acceptance or rejection and use statistical techniques to see which ones tend to cluster together in people's minds."

Recognizing that many respondents might not support the space program, and indeed relying upon such negativists for comparison with space-boosters, I also wrote: "Whatever your feelings about the space program, we hope you will share them by completing this questionnaire. While most of the items in this questionnaire are stated in an apparently 'pro-space' manner, there is ample opportunity for you to express negative sentiments, if you have them."

Then followed the 125 space goals, each with a scale of seven numbers, zero through 6, for students' responses. The instructions called them *justifications for the space program,* and said: "You will probably feel that some reasons are much better than others. Don't worry about all the aspects of each one, but make an overall judgment of it. After each statement is a scale from '0' (not a good reason) to '6' (an extremely good reason). For each justification, please circle the ONE number that expresses how good a reason you feel it is for supporting the space program."

The 125 space goals were listed in random order—indeed in five different random orders. As in the pilot study, I was concerned that the mere placement of an item in the survey might distort responses. Appendix B of this book lists them in the first of my five random orders, each bearing the identifying number I will use in other tables and occasionally in the text.

After the space goals, the survey contained twenty-two miscellaneous questions, the number limited by the already great length of the survey. Several of these will help us make sense of the long list of goals, beginning later in this chapter, and others provide brief descriptions of the respondents. Sixty-five percent were men, and 35 percent women. Roughly equal numbers were freshmen (22.8 percent), sophomores (18.8 percent), juniors (29.0 percent), and seniors (24.2 percent). Just 3.6 percent were graduate students, and 1.7 percent were in the inevitable "other" category. Fully 93.2 percent were citizens of the United States. When asked to mark one of eight general academic fields as their favorite, respondents revealed great diversity of interests: arts (7.4 percent), humanities (26.7 percent), social sciences (30.1 percent), biological sciences (15.8 percent), physical sciences (13.5 percent), and mathematics (6.5 percent). Another miscellaneous item inquired, "In terms of general factual knowledge, how much do you know about the American space program?" Only 2.0 percent of the Harvard students admitted to knowing "nothing," and 24.2 percent said "very little." The majority, 57.0 percent, said "a moderate amount," and 16.8 percent knew "quite a lot."

AN OVERVIEW OF SPACE GOALS

We can begin to get familiar with the 125 space goals by looking at the ones Harvard students ranked highest. Remember that our data are designed for exploring the concepts on which the goals rest, not for predicting the attitudes toward them held by a cross section of the American public. However, Harvard students are an interesting segment of our future leaders, and it will help us begin analyzing the goals of the space program by inspecting the ones for which they expressed the most enthusiasm. Of the 125, 17 achieved average ratings of 4.00 on the 0 to 6 scale, and they are listed in Table 3.1, along with their mean ratings and the percent of respondents giving them a 5 or 6 rating.

Table 3.1. Goals Judged 4.00 or Better by 894 Harvard Students.

	Mean Rating	Percent 5 or 6
31. The space program contributes to the advancement of science.	4.46	54.9
12. New medicines could be manufactured in the zero gravity and vacuum of space.	4.37	53.0
79. Medical research performed in space could benefit human health.	4.35	53.8
58. Space probes increase our knowledge of space, planets, comets, and the entire solar system.	4.31	51.2
44. Meteorology satellites are great aids for predicting the weather and understanding atmospheric patterns.	4.25	46.1
9. The long-term, ultimate benefits of the space program could eventually be important.	4.24	52.9
63. Solar power stations in orbit could provide clean, limitless energy to the Earth.	4.24	51.0
19. Satellites link all corners of the globe in a complete information and communication network.	4.24	49.8
49. Many experiments can be done best in the environment of space.	4.20	49.2
98. In the weightlessness and vacuum of space, we could manufacture new and better alloys, crystals, chemicals, and machine parts.	4.18	47.7

Table 3.1. *(continued)*

	Mean Rating	*Percent 5 or 6*
38. Space research tests our scientific theories and promises conceptual breakthroughs.	4.18	47.3
15. The space program contributes much to our technology.	4.11	45.8
89. Space research provides valuable, practical information.	4.09	44.1
78. Space research benefits physics— in studies of the nature of matter, for example.	4.08	41.7
113. Space could offer many unexpected benefits we cannot now foresee.	4.05	46.3
72. New fuels found in space or the development of fusion power in space could help solve the Earth's energy problem.	4.03	44.9
88. Satellite photography of the Earth contributes to geology, oceanography, and archaeology.	4.00	38.3

The very highest rating goes to a general statement of space-flight's scientific goals: "The space program contributes to the advancement of science." Apparently respondents are convinced that the space program does in fact contribute to the advancement of science, and they feel this is a very important benefit.

Four other items in the top seventeen also talk about science: "Space probes increase our knowledge of space, planets, comets, and the entire solar system." "Many experiments can be done best in the environment of space." "Space research tests our scientific theories and promises conceptual breakthroughs." "Space research benefits physics—in studies of the nature of matter, for example."

Second and third in the ranking come two items about advances in medicine that might come from the space program: "New medicines could be manufactured in the zero gravity and vacuum of space." "Medical research performed in space could benefit human health." Thus, respondents see space not merely as an extension of the so-called hard sciences, astronomy and physics, but as having value for biological sciences and medicine.

Respondents note the important function orbiting satellites have in monitoring our world: "Meteorology satellites are great aids for predicting the weather and understanding atmospheric patterns." "Satellite photography of the Earth contributes to geology, oceanography, and archaeology." Perhaps most of them are aware of the substantial benefits the space program has already achieved in this area. They also give a high rating to "Satellites link all corners of the globe in a complete information and communication network." Stating the point in more general terms: "The space program contributes much to our technology." "Space research provides valuable, practical information."

Three goals for the near future concern power generation and manufacturing in space. "Solar power stations in orbit could provide clean, limitless energy to the Earth." "New fuels found in space or the development of fusion power in space could help solve the Earth's energy problem." "In the weightlessness and vacuum of space, we could manufacture new and better alloys, crystals, chemicals, and machine parts."

But immediate benefits are not the only goals of the space program, and it is a sign of respondents' faith in the future that they give high ratings to two items of vast scope: "The long-term, ultimate benefits of the space program could eventually be important." "Space could offer many unexpected benefits we cannot now foresee."

On the face of them, these seventeen high-rated goals of spaceflight are entirely reasonable, and the reader may agree that our 894 respondents have shown good taste in selecting them. Other goals are more controversial, and some draw almost universal rejection. Of the 125 possible goals, 7 got average ratings below 1.5, and on 5 of these a majority of respondents expressed the lowest possible zero rating.

Three of the lowest-ranked goals can be described as frivolous and far too unimportant to justify the large expenditures required for a vigorous space program. "Vacations and games in space could be entertaining." "I like watching rockets take off and enjoy space probes." "Jokes and cartoons about the space program can be amusing." Certainly science fiction stories have often suggested that the idle rich of future centuries may vacation and play in space, but today's citizens have little reason to subsidize such frivolity. Personally, I do enjoy watching rockets take off, from the tiny ones that lifted only a few dozen feet from my back yard in youth to the Apollo 17 that last carried men to the Moon and colored the Florida night sky orange with its fire in 1972, but taxpayers have no obligation to support my personal pleasures. In the wake of the Challenger explosion, a series of sick "Christa McCauliffe" jokes swept the country, and our respondents rightly place their ilk at the absolute bottom of list of space goals.

Two other rejected goals suggest that certain deviants be exiled from our planet. "The worst criminals could be put in space prisons." "AIDS victims could live safely in the complete isolation from infection that can be assured in space." Science fiction writers have often speculated that among the first colonists of the Moon and planets might be convicts, perhaps political prisoners, and many Australians are descended from convicts transported to their island continent a century and more ago. At least in principle, victims of Acquired Immune Deficiency Syndrome (AIDS) could avoid many of the devastating consequences of their illness if their environment were entirely free of disease-producing organisms, something one can conceive of achieving in outer space. But our respondents find little merit in these ideas and probably see them as springing from bigotry and arrogance rather than from humane consideration of the related social problems.

The two remaining low-rated items might get considerably more favorable responses from other groups of respondents. Sensitive to the traditions of other cultures, Harvard students probably see distasteful jingoism in the statement that "the space program represents the best traditions of Western Civilization." Only three S1986A respondents expressed ideas along these lines, but that was enough to get the idea into the list of possible goals.

Although many students possess a private faith in God, and religion thrives both in the Harvard Divinity School and various campus ministries, secularism reigns supreme in the university at large. Respondents may never have seen any of the theological arguments about religious implications of space exploration, and it would be interesting to investigate the topic further with people who had given them close consideration. Only two S1986A respondents saw religious goals for the space program, defining a sacred goal: "Space discovery has religious implications, leading us to God."

Setting aside the debates one could have over this final pair of items, our respondents show good sense in the space goals they reject, as well as in those they rate highly. One can test how reasonable our respondents are also by inspecting the correlations linking goals with each other. The strongest correlation is 0.72 between two items that look almost synonymous: "There are great military applications of space." "The space program contributes to our defense." When I was collating the S1986A responses, I thought carefully about this pair. Some students had written about "defense" and others about "military" matters, and I was strongly tempted to place all their utterances in the same pile and make them into just one goal statement for S1986B. But it seemed to me that "defense" had a positive connotation, whereas "military" might have a negative connotation, as Harvard students used the term, so I split these two sets of utterances into two goals. But, respondents to S1986B put them back together again.

The second highest correlation (0.71) links two highly idealistic items that express pride in the human species. "Spaceflight reaffirms faith in man's abilities." "Space exploration is a human struggle, expressing the unconquerable human spirit." The third highest correlation (0.69) links the latter of these two with another of the same type. "Spaceflight is a noble endeavor, expressing the hopes and aspirations of humankind." Perhaps I should have combined two or three of these items in the survey, and it is certainly reasonable for the students to connect them. However, respondents could not possibly keep all 125 goals in mind at once, and few may have had any conscious awareness that these specific items were similar. The high correlations that emerged in the statistical analysis show that respondents react with a fairly consistent set of interpretations and concepts from which to judge.

I had the computer locate the highest correlation for each item, and these are listed in Appendix B. Only four of the goals failed to achieve a correlation of 0.40 or greater with at least one other goal. Three were already listed among the seven with the worst ratings, the religious implications item and the pair about criminals and AIDS victims.

The remaining lost item was probably meant as a joke by the S1986A respondents who mentioned it. "Sex in the zero gravity of space could be quite an experience." Apparently 282 autumn 1986 respondents liked the joke and gave the item a 6 rating, whereas 375 gave it a 0, leaving only 26.5 percent of the respondents to give all the five scores between these extremes. By far, this is the item getting the most extreme scores. Frankly, this item worried me when it emerged out of the S1986A utterances. My research method rules required me to include any idea expressed clearly by two or more respondents, and this one just met the criterion with three mentioners. One of them frivolously proposed not only "zero gravity sex," but also "asteroid rodeos" and "artistic repositioning of the planets!"

The fact that I made five editions of S1986B, with the 125 goals in five different random orders, kept this item, or any other items that receive extreme responses, from contaminating responses to neighboring items. And in the final statistical results, this absurd item drops out. Its highest correlations (about 0.35) are with the junk items about vacations in space and space jokes. Undoubtedly, this is the item eliciting the weirdest responses, and yet the statistical result is completely reasonable.

Thus, the respondents had no difficulty separating the bad or poorly conceived justifications for the space program from the much larger number of valid space goals. Even as we cringe at some of the foolish things that may be said about the value of spaceflight, we can take renewed confidence in both the respondents and the research methods of this research study.

Altogether, there are 7,750 different pairs of items among the 125, thus 7,750 unique correlation coefficients, and the average of all these is 0.26. Generally, we will not be interested in correlations this low, especially between different space goals. One reason is that the psychological phenomenon called *response bias* tends to produce a low positive correlation linking all the questionnaire items of any given format (Block 1965). Some people tend to express more enthusiasm about everything than do some other respondents. Some tend to agree with items in a survey, out of politeness or deference to the authority of the printed word, whereas others tend to say whatever they think is socially most acceptable. Usually, these response biases are mild, and we can guard against them here by trusting only the biggest correlations and basing much of our analysis on comparisons between correlations of different magnitudes.

Another factor that produces a background positive correlation between all the goals is the fact that each is a measure of the respondent's general attitude toward the space program. Persons who favor the space program will tend to rate each of the goals higher than people who disfavor it. We can document this through another question in the survey, a space funding item taken from a standard national poll, and at the same time locate the goals that most clearly reflect the perspective of enthusiasts for the space program.

CENTRAL BELIEFS OF SPACEFLIGHT ENTHUSIASTS

As I explained earlier, a set of questions included in the General Social Survey (Davis and Smith 1986) concerns appropriations for government programs, and I included them in S1986B. The students interested enough to fill out our survey expressed strong support for the space program, 41.3 percent of them saying "too little" was being spent on it, and 38.5 percent felt current appropriations are "about right." Only 9.8 percent believe too much is being spent on space, and 10.1 percent "don't know." Those saying "about right" and "don't know" can be combined for purposes of correlational analysis. Keep in mind that our research method does not employ a random sample, and it is not primarily devoted to estimating what percentage of Americans, college students, or even Harvard undergraduates wants space funding increased. However, highly educated populations tend to give high levels of support, so these percentages are not surprising.

To explore the extent that attitudes toward the 125 goals partly represent general support for the space program, I examined correlations linking each of them with this space funding item. On average, there was a modest association (Pearson's $r = 0.23$) between rating a goal high and wanting appropriations increased. This does not mean that all the correlations in the set are inflated by this amount. For one thing, when comparing correlation coefficients statisticians like to square the numbers first (to get "explained variance"). That is, a correlation of 0.50 is not twice as big as one of 0.25, but really four times as big in terms of the variation explained by it.

Aside from its help in determining that the background positive correlation is small, this space funding item can help us single out goals that are somehow central to the spaceflight creed. That is, we can ask which goals get an especially favorable response from respondents who want funding increased— which have the biggest correlations with funding. It turns out that six goals are tied for first place, having the same 0.36 correlation with space funding. In a sense, the six are the heart of the spaceflight ideology, because they are the goals that best distinguish supporters of the space program from nonsupporters.

Three are expressions of very general faith in the spaceflight enterprise. "The long-term, ultimate benefits of the space program could eventually be important." "The space program contributes much to our technology." "Our future ultimately lies in space." The other three look outward from our Earth to a future age in which humans build an interplanetary civilization and contemplate the vastness beyond. "We could establish manned space stations, communities in space, and space cities." "We could colonize the moon, Mars, and other satellites or planets of our solar system." "An orbiting space telescope could give astronomers a much better view of the stars." Although this last space goal looks toward a limitless future of exploration, it is well within our grasp today.

BLOCK MODELING THE GOALS

In the previous chapter, we used factor analysis to find underlying conceptual dimensions that connected many of the forty-nine space goals we had identified. Factor analysis demands a lot from computers, and no statistical package available to me was able to handle a factor analysis with 125 variables. However, I had been experimenting with alternatives to factor analysis, various kinds of so-called cluster analysis (Everitt 1974), and eventually wrote my own programs for working with data from S1986B and finding clusters of goals that correlated highly with each other.

But the decision to abandon factor analysis had good methodological reasons, as well as practical ones. Factor analysis hunts for dimensions that run through all the items, whereas cluster analysis may be better at finding separate groups of items. Typically, factor analysis concentrates on major dimensions of variation, letting lesser clusters drop through the cracks. I wanted a technique that would stay close to the correlations themselves, helping us understand the link between goals without submerging them in excessive abstraction.

Furthermore, one reason for using a different approach in the main study, paradoxically, was that the pilot study had succeeded so well. I wanted to chart new territory. By using a different statistical technique I could achieve fresh insight and, at least, subject the original findings to a good test. Although details are bound to differ, if the new analysis gives results comparable to those from the factor analysis, then I will have achieved a solid replication of earlier findings.

The technique I chose and adapted is called *block modeling*. Long ago, social psychologists used this method for finding groups of friends in networks of relationships between individual respondents (Festinger, Shachter and Back 1950). A generation later, Harrison White and his associates updated block modeling by computerizing it and providing a sounder mathematical basis (White, Boorman, and Breiger 1976). Recently, I adapted it to analysis of a correlation matrix. My laboratory workbook and software module, *Experiments in Psychology* (1985), and my software-textbook package, *Survey Research* (1989a, 1989b), show visually how it works.

We start with the entire correlation matrix, a grid 125 items on a side, including 15,625 squares, each with a correlation coefficient inside it. First, the software user decides how big a coefficient has to be before it is worth including in the analysis, setting the *threshold criterion*, as we might call it. On a given computer run we might be interested only in correlations 0.50 or greater, for example. If so, 0.50 is our threshold criterion. The computer will go through the matrix, inspecting each of the 15,625 coefficients. Any coefficient less than 0.50 will be turned into a 0. A coefficient that achieves the criterion, being 0.50 or greater, is turned into a 1. The matrix becomes filled with 1s and 0s.

Now the hard part begins for the computer. Over a period of many hours, it fiddles with the order in which the 125 space goals are listed, trying to switch them around to bring the ones in the matrix closer together. Success means getting the 1s into nice, neat blocks representing high correlations linking goals in particular clusters. The fiddling is done automatically, with no help from the researcher. Again and again, the computer selects two goals at random. It calculates how the matrix would change if the positions of the two goals were switched in the list. If switching them would bring 1s (representing big correlations) closer together, the computer makes the switch; if not, it doesn't.

After 50,000 attempts to switch goals in the list, and 200 to 500 actual switches, the computer typically ran out of changes to make. I always double checked this with a related program that determined whether any more switches could be made. Now, unlike factor analysis, block modeling does not give precisely unique results; running a second time might place the items in a slightly different final order. Also, the process can get stuck, arriving at an order that could be improved if three or four items could be switched simultaneously, rather than just two. But there is a tool for overcoming these limitations—the human mind.

When the computer was finished with each run, I had it store its final arrangement of the correlation matrix on disk, then print it out as a huge map of 1s and 0s. This is the *block model* itself, a perspective on the structure of correlations linking all the items. It is very easy then to spot any anomalies on the block model, because they are represented by a few lone 1s scattered far from the others.

After fine tuning the procedure and checking initial results, a combination of three block models gave excellent guidance for understanding the structure of spaceflight values, the runs that analyzed the correlation matrix at three threshold criteria—correlations of 0.50, 0.45, and 0.40. Done entirely independently, they give very similar pictures. The 0.50 criterion shows us just the most important connections, including the hearts of the main clusters of correlations. The 0.40 criterion takes us down to the level at which substantial numbers of items are combined into clusters. And the 0.45 criterion provides an intermediary check on the other two.

Like factor analysis, block modeling can produce different numbers of clusters, depending on exactly which commands we give to the computer. The point is not to settle on the one, perfectly valid analysis—there is no single one. Rather, the point is to find a level of analysis that suits our need for understanding, balancing lumping with splitting to give us the clearest and most useful picture for our particular purposes. Because my purpose was communicating results through a book, I wanted an analysis that, first of all, would allow me to separate sets of goals into different chapters. Then, within each chapter I could engage in further splitting, examining subclusters of goals and bridges that linked one cluster with another.

The block modeling analysis did, indeed, replicate the main contours of the factor analysis from the pilot study. One, very clear small cluster consisted of six military or international competition goals, essentially identical to the Military factor in the previous chapter. Along with other, related material, it is the focus of Chapter 6. Also rather clear was a larger group of space goals having to do with the colonization of space, from the intensive exploitation of near-Earth space to the actual planting of human society on the moons and other planets of the solar system. This Colonization class of goals is the subject of Chapter 7. Communication with extraterrestrial intelligence did not feature as a major aspect of the space program in responses to questionnaire S1986A, and only a couple of space goals were oriented toward it. However, it is an interesting extension of space development, and I have quite a lot of data about it, so Chapter 8 will examine this subject.

Two other major clusters developed in the block modeling analysis, each comprising a large number of specific space goals. One is essentially identical to the Idealistic-Emotional factor discussed in the previous chapter, and it will be the topic of Chapter 5. The other combines purely scientific goals of spaceflight with the industrial and economic goals, thus placing in one class two of the factors found in the pilot study. Because of this small shift, we will give it a fresh name, the Technical class of goals. In Chapter 4 we will examine this group, contemplating the extent to which students mentally link science with technology, technology with industry, and industry with economy, to see these diverse goals as parts of a whole. A satellite cluster, very loosely connected to science-industry, consisted of four statements about protecting the environment from pollution, and they also are analyzed in Chapter 4.

A few goals were pulled equally toward one or another cluster, two items about learning the origin of life and the history of our world acted as a bridge between the Technical and Idealistic classes. Other goals span from the Idealistic class to the one concerning space colonization, but I found it possible to assign them primarily to one or the other. Finally, four loose items failed to correlate highly enough to participate in the 0.40 block modeling, space junk we had best forget.

I must again caution that other sets of respondents would rate particular goals or clusters higher or lower than do Harvard students, but the average ratings of goals in these clusters are worth comparing, as done in Table 3.2. Like the Seattle voters, Harvard students place the scientific and economic class of goals highest, believing that these intellectual and practical benefits are the best justifications for a vigorous space program. Lowest they place the Military class, a reflection of the political liberalism of their campus.

Table 3.2. Average Rating of Classes of Space Goals

Cluster (Block)	Number of Items	Mean Rating
1. Technical	32	3.87
Environmental Satellite	4	3.23
Origins Satellite	2	2.94
2. Idealistic	55	2.42
3. Military	6	2.07
4. Colonization	20	2.69
5. Communication with Extraterrestrials	2	2.65
Loose items	4	1.35

The full utility of the block models, and the meaning of the clusters of goals themselves, will be developed in the following chapters, beginning with the most highly rated justifications for the space program, those identifying the scientific knowledge, technological advances, and economic benefits that have already begun to flow from it.

CHAPTER FOUR

TECHNOLOGY, SCIENCE, AND ECONOMICS

On its journey to the stars, mankind has found many ways in which exploitation of space can directly benefit our home planet. From the earliest years of the space age, satellites contributed to terrestrial communications and weather prediction. The very first American satellite achieved scientific discoveries, and our knowledge of the universe around us has expanded considerably through observations above the atmosphere and probes to other celestial bodies. When Americans consider the goals of the space program, they value most the technical and intellectual gains that it has already achieved and they expect further benefits of a practical kind. Spaceflight, the creation of a social revolution, has become part of normal technological progress.

The class of goals we shall consider in this chapter contains thirty-two questionnaire items about the technological, scientific, and economic goals that the space program has already substantially achieved. One way to evaluate the distinctness of this group is to look at the average correlations linking items in it and compare them with the average to items that are not members of the class. The mean intraclass correlation is 0.39; that is, the average correlation of each of the thirty-two member items with each of the other thirty-one is 0.39. In contrast, the average correlation connecting each of the thirty-two with the ninety-three items that are not members of the class is 0.22. All the items measure support for the space program to some extent, so one would expect them all to be associated. Classes stand out because their internal links are especially strong. The ratio of average intra-class coefficient to extraclass coefficient is 0.39 divided by 0.22, or 1.77. However, statisticians usually like to square correlation coefficients (to get proportion of variance explained) before comparing them. The ratio of the squares is 3.14.

Average intraclass correlations can also help us understand the root concept that defines the category. Individual items have average correlations with the other thirty-one ranging from 0.29 to 0.45. Twelve achieve mean coefficients of 0.40 or greater, and we can take them to represent the core of the entire cluster. The highest correlation belongs to a practical but future-oriented goal: "In the weightlessness and vacuum of space, we could manufacture new and better alloys, crystals, chemicals, and machine parts." Second, with 0.44, is "Techno-logical spin-offs (advancements developed for the space program, then applied to other fields) improve everyday life."

Six goals are tied at 0.43: "Satellite photography of the Earth contributes to geology, oceanography, and archaeology." "Satellites are an important component in navigation systems." "The space program contributes much to our technology." "Space research provides valuable, practical information." "The space program has great benefits for industry." "Space research benefits physics—in studies of the nature of matter, for example."

Two goals are tied at 0.42: "Satellites are useful in surveying and mapping the Earth." "Medical research performed in space could benefit human health." Another pair is tied at 0.40: "Space research tests our scientific theories and promises conceptual breakthroughs." "The space program contributes to the advancement of science." Clearly, this class of goals links science, industry, and technology. Already, in the slight differences of average correlation, we have a hint that the technological element predominates over the scientific. With full acknowledgment that scientific knowledge and industrial applications are integral aspects of the cluster, for convenience I shall refer to it as the Technical class of goals.

Within this group of thirty-two are subgroups, and we need both terminology and methodology to distinguish them. I borrow my hierarchical scheme of category nomenclature from biology. The utterances collected in S1986A are like individual organisms, having unique characteristics as well as shared traits. Specific concepts, often expressed by several individual utterances, are at the level of *species* in this typology. The collections of utterances that produced the 125 goal statements are at a higher level of abstraction, however. Often, species of utterances describe aspects of a goal. For example, the species of utterances about satellite transmission of telephone calls contributes to the same goal as the species of utterances about long-range computer data transmission via satellite. Thus, our 125 goals are at the level comparable to the biological concept of *genus*. Each goal is a genus of justifications for the space program. Each utterance could be efficiently described in terms of its genus and species; for example, genus communications satellites, species telephone transmissions.

Above genus, in ascending order, come family, order, and class. I will use the word *family* for a small set of goals that are very closely connected, for example, defense and military. An *order* of goals is a group that is both statistically and conceptually distinguishable, although usually not so different from other goals as to require a separate chapter in this book. A *class* of goals does require its own chapter, unless it is so ambiguous or insignificant that the best we can manage is a section tacked onto the end of a chapter. Thus, the thirty-two questionnaire items that are the prime focus of this chapter are the *class* of Technical goals.

Because the class has fully thirty-two items, we will need a systematic approach to make sense of all the information contained within it. I found that a clear structure appeared in the cluster if I simply worked with the largest correlation each goal has with other items in the list. This is a standard technique in cluster analysis (Everitt 1974), and although it is rather sensitive to small difference in correlation, the results were so clear for the goals considered in this this chapter and the next, that I feel safe in using it.

I tabulated the items most strongly connected with each of the thirty-two, finding only nineteen different ones, and all were fellow members of the cluster. I drew simple charts of the items connected by highest correlations, finding eight subclusters with the following numbers of items: 9, 6, 5, 3, 3, 2, 2, and 2. We shall consider each of these *orders* of goals, beginning with the largest.

THE EARTH-COMSAT ORDER

A group of nine Technical items, is headed by two items and might be divided into two connected families of goals. The pair correlate at 0.64: "Satellites are useful in surveying and mapping the Earth." "Satellites are an important component in navigation systems."

Three other goals have their highest correlations with satellite mapping: "Observations from orbit help us find new sources of energy and minerals on the Earth." "Meteorology satellites are great aids for predicting the weather and understanding atmospheric patterns." "Satellite photography of the Earth contributes to geology, oceanography, and archaeology." Another item has its highest correlation with the last in this list: "We could understand the Earth's atmosphere and geology better, through comparisons with other planets." Clearly, these five concern gaining information about the Earth, especially its atmosphere and geology.

Three other items have their highest correlations with navigation satellites: "Communication satellites improve television transmissions." "Satellites link all corners of the globe in a complete information and communication network." "We must be able to launch, retrieve, and repair satellites." Two of the four in this group talk about communication satellites (comsats), and the navigation satellites are also engaged in communication, if of a very special kind, so we can refer to this group as Comsat goals.

In fact, most of the nine items are strongly connected to each other, and the two subgroups are really *families* within the same order. The average of the thirty-six linkages between these nine items is 0.45. Within the Earth family it is 0.51; within the Communications family it is 0.45; and between the two the average correlation is 0.42. Thus, the division of the order is shallow, and it can be considered a unit, the Earth-Comsat order. Table 4.1

Table 4.1. The Earth-Comsat Order (correlations)

	Average Correlation		68	88	75
	Intra-class	Extra-class	Survey Earth	Photo Earth	Navi-gation
Earth Observation Family:					
68. Satellites are useful in surveying and mapping the Earth.	0.42	0.19	1.00	0.62	0.64
23. Observations from orbit help us find new sources of energy and minerals on the Earth.	0.38	0.19	0.51	0.50	0.45
44. Meteorology satellites are great aids for predicting the weather and understanding atmospheric patterns.	0.38	0.15	0.56	0.52	0.54
88. Satellite photography of the Earth contributes to geology, oceanography, and archaeology.	0.43	0.20	0.62	1.00	0.59
10. We could understand the Earth's atmosphere and geology better, through comparisons with other planets.	0.39	0.21	0.47	0.57	0.45
Communication Satellite Family:					
75. Satellites are an important component in navigation systems.	0.43	0.20	0.64	0.59	1.00
119. Communication satellites improve television transmissions.	0.29	0.18	0.37	0.33	0.50
19. Satellites link all corners of the globe in a complete information and communication network.	0.36	0.17	0.50	0.44	0.53
117. We must be able to launch, retrieve, and repair satellites.	0.36	0.19	0.43	0.41	0.50

shows average correlations linking items with others in the class, average correlations with the other ninety-three goals, and correlations with the three goals having the highest average correlations with fellow members of the order. "Satellites are useful in surveying and mapping the Earth" has a correlation of 0.52 with the eight other goals in the order. The item about satellite photography achieves 0.50, ands the one about navigation gets 0.53.

Despite the clarity of this order and the popularity of its goals as measured by survey S1986B, the open-ended statements from survey S1986A did not express the ideas in very great detail. Goal 68, the first in Table 4.1, came from a mere three utterances: "surveying," "advanced mapping and search techniques," and "satellite mapping." The Earth resources goal was stated in a similarly sketchy manner: "Better resource searching," "mineral exploration benefits," "the discovery of vast fossil fuel reserves." One utterance refused to limit itself to terrestrial resources: "Perhaps searching for new mineral resources both on Earth (by means of satellites) and elsewhere." Three others spoke of "discovery of new energy resources" without conclusively stating that they were to be found on this planet.

Every respondent to S1986A must have known about meteorology satellites, but the twenty-one utterances that mentioned them gave no details beyond "weather information," "weather prediction," or "increase in meteorological knowledge." The satellite photography goal was based on statements about "Earth imaging," "photography for geological reasons," and the like. Photographic satellites "increase Earth information by looking from space," achieving "data collection for oceanographic research" and "research into plate tectonics." "They take great photos that help archaeologists."

Space goal 10 apparently takes us far from the practical concerns that dominate this chapter, investigating conditions on other planets, but the purpose of that far-out research is achieving a better understanding of our home world. Consider "the knowledge we gain about our own Earth's tectonic and weather systems by studying other planets." "Studying planets and the outer atmosphere can better help us understand the environment of Earth (e.g., nuclear winter and the climate of Venus)." Space research "can tell us much about our planet and our system by exploring others."

The navigation goal was expressed in only three, terse statements: "navigation satellites," "navigation satellites," "navigation systems." Perhaps I should have combined goals 119 and 19, because both concern communication satellites, but the former did seem to focus more narrowly on television transmission, whereas the latter covers the range of communication technologies and their social implications. "TV satellites" certainly provide "better TV programs and reception." "Satellites cheapen communication" and "speed up telecommunications." "Expanding communications capability affects all American life" and "has helped international communication." Here space applications shade over into spin-offs. "Communication technology can certainly benefit from the space program," and "development of much of today's vital telecommunications comes directly from NASA research."

Because S1986A was taken right after the Challenger disaster, some respondents had given thought to the difficulties we face without the capability of launching and repairing satellites. "Satellite repair and launch and retrieval" are vital, and "maintenance and positions of satellites are very useful."

THE SPIN-OFF TECHNOLOGY ORDER

One often hears the word *spin-off* applied to any direct, practical benefit of the space program, but it is best to distinguish spin-offs from applications. As the first goal in the order defines the concept, *spin-offs* are "advancements developed for the space program, then applied to other fields." Thus they are examples of technology transfer, often serendipitous, that finds unintended value in scientific or technological developments made in the space program. In contrast, a space *application* is an intentional use of space, such as those listed in Table 4.1. The six goals of the spin-off technology order are listed at the top of Table 4.2.

The central goal is, of course, the concept of spin-offs itself. It achieves the highest correlation with fellow goals in its order, 0.52. Closely related is a specific area of spin-off: better computers, calculators, and electronics. Advances in high-speed aircraft could represent an even more specific spin-off, and respondents focusing on the goals of the space program may not recall that the first *A* in NASA stands for Aeronautics.

Table 4.2. The Spin-off Technology Order (correlations)

	Average Correlation		33	15 Contributes to Technology	53 Improves Life on Earth
	Intra-class	*Extra-class*	*Spin-offs*		
33. Technological spin-offs (advancements developed for the space program, then applied to other fields) improve every-day life.	0.44	0.24	1.00	0.58	0.42
60. The space program produces better computers, calculators, and electronics.	0.39	0.22	0.56	0.54	0.37
54. Our ability to travel on Earth could be improved —with high-speed aircraft, for example.	0.35	0.23	0.46	0.37	0.29
15. The space program contributes much to our technology.	0.43	0.26	0.58	1.00	0.49
89. Space research provides valuable, practical information.	0.43	0.23	0.50	0.55	0.49
53. Space could improve life on Earth, providing benefits to mankind.	0.37	0.25	0.42	0.49	1.00

Related Goals:

31. The space program contributes to the advancement of science.	—	—	0.55	0.54	0.37
38. Space research tests our scientific theories and promises conceptual breakthroughs.	—	—	0.50	0.46	0.30
97. The space program has great benefits for industry.	—	—	0.54	0.56	0.41
98. In the weightlessness and					

vacuum of space, we could manufacture new and better alloys, crystals, chemicals, and machine parts.	—	—	0.53	0.50	0.39
106. Space has great commercial applications and many opportunities for business.	—	—	0.51	0.51	0.41

The concept of spin-offs is implied by the statement that, "The space program contributes much to our technology." Indeed, this item has the second-highest correlation with items in the order, 0.50. In its final items, the order shades from narrowly technological benefits to the more general concepts of practical information, improving life on Earth, and benefits to mankind. Thus, respondents mentally connect general ideas of human benefit to the technological spin-offs from space development.

Table 4.2 shows correlations between all the order's goals and three of them: spin-offs, contribution to technology, and improving life on Earth. As proof of their centrality to the order, the first two of these achieve the highest correlations. The lower correlations possessed by the goal of improving life on Earth probably come from the fact that respondents can think of very different possible improvements than those associated with technological spin-offs. Thus, it is so general a goal that it cannot achieve the high coefficients gained by spin-offs and technology. At the other extreme, however, a goal can be so narrow that its correlations with more general items are somewhat weak; for example, the goal of high-speed aircraft. This comparison shows that the scope of an item—the range of human experience and activities it covers—shapes its correlations. Other things being equal, two goals of similar scope can achieve a higher correlation than two of very different scope, assuming of course that the two actually are related in respondents' minds.

The bottom of Table 4.2 shows five goals that are not members of the order but achieved correlations of 0.50 with the spin-off or technology item. The strong correlations with the advancement of science are noteworthy. Throughout this chapter, we will continually see science connected to technology, although some scholars have gone to great lengths to distinguish them (Fores 1969), and each can develop somewhat separately from the other (Jewkes, Sawers, and Stillerman 1969). This connection says much about our respondents' conceptualization of science, indicating that they are interested in the practical consequences of science perhaps more than the abstractions of theory. The three related goals at the bottom of the table concern industry, manufacturing, and commercial applications, showing the links between the spin-off order and the other practical goals of the class.

In the previous chapter, we discussed the utterances from S1986A that generated the spin-off item. The remaining five goals in the order can be covered quickly. The "practical improvement in electronics" primarily meant "advanced computers" and "calculators," with a couple of mentions of "better silicon chips" and "solar cell development." "Advances in transportation" on Earth might mean "hyper-speed planes" or a "Washington to Tokyo shuttle."

Thirteen utterances simply said "technology," and 140 others used the word in some form: for example *technological progress*, *technological advancement*, *new technologies*, *technological benefits*, *technological breakthroughs*, *technological drive*, and *techno-boom*. "Technology development has always resulted from space exploration," "as evidenced by 1960s race to the moon." "Techno advances equal economic advances," producing "better living standards" and a "better lifestyle for all." The space program is a "testing ground for new technology," stimulating "private sector tech innovation." "It's the future of technology." "It's the ultimate achievement of our technology."

Respondents favored the space program "because of the vast knowledge, theoretical and practical, that we could possibly gain." "A good space program adds to humanity's general knowledge, and with the more knowledge we have, our lives would inevitably be better." "The gains made by research efforts" are "spread to all people, because NASA doesn't keep secrets." "Knowledge is power," and "knowledge, of any sort, is the most useful commodity." In space the "pursuit of knowledge" will "help mankind," will "discover new things that may help the world."

In a general way, they also felt that "learning about space

and how to use it for our benefit should be an important goal for us." Space should be used "for the benefit of mankind," "to improve life on Earth" and "achieve gains for humanity." "We should continue to explore space for answers to our problems on Earth." "We can possibly conquer scarcity on Earth," In this tone they stated the goals for a good space program: "To learn about our universe, but not to exploit it for evil and destructive ends. To gain knowledge that we can use to better our world." "To make the world a better and safer place to live. To raise the quality of life of every human being in the world."

These ideals do not place the Spin-Off order in the Idealistic class. On average, its six goals correlate 0.40 with fellow members of the Technical class, compared with 0.24 to the 93 extra-class goals. Connections within the order are generally stronger than those to other orders in the class. For example, the average intraorder coefficient is 0.45, compared with 0.37 connecting the items in this order with those in the Earth-Comsat order. The average correlation across orders, among the first four discussed here, is 0.39. Thus, the Technical class has high solidarity, and the division into orders is a relatively weak one.

THE EXPLOITATION OF
SPACE CONDITIONS ORDER

The next order might deserve the name *medical* or *biological*, because three of its five items concern these topics. However, some life-related items are not included here, and the remaining pair of goals in the order have nothing to do with medicine. All five identify benefits of the unique natural conditions of space itself: zero gravity and high vacuum. The first part of Table 4.3 lists the five goals concerning exploitation of orbital conditions, with correlations to three of them. The goal with the highest average correlation to fellow members of its order is actually number 79, about medical research, achieving 0.61. But the divisions within the order are revealed slightly better by using those in second, third, and fourth place, with insignificantly smaller intraorder correlations of 0.60, 0.58, and 0.58.

Table 4.3. The Exploitation of Space Conditions Order

	Average Correlation		12	99	98
	Intra-class	Extra-class	New Medicines	Medical Treatment	Manu-facture
79. Medical research performed in space could benefit human health.	0.42	0.21	0.68	0.68	0.59
12. New medicines could be manufactured in the zero gravity and vacuum of space.	0.39	0.18	1.00	0.63	0.60
99. Some medical problems could be treated more effectively in the weightlessness of space.	0.38	0.22	0.63	1.00	0.55
98. In the weightlessness and vacuum of space, we could manufacture new and better alloys, crystals, chemicals, and machine parts.	0.45	0.20	0.60	0.55	1.00
49. Many experiments can be done best in the environment of space.	0.38	0.18	0.48	0.49	0.56
Related Goals:					
15. The space program contributes much to our technology.	—	—	0.41	0.37	0.50
33. Technological spin-offs (advancements developed for the space program, then applied to other fields) improve every-day life.	—	—	0.41	0.42	0.53
78. Space research benefits physics—in studies of the nature of matter, for example.	—	—	0.42	0.42	0.51
97. The space program has great benefits for industry.	—	—	0.37	0.41	0.51

The average correlation within the order is 0.58, indicating high cohesion among the five goals. The three medical items make a tight family, sharing extremely high correlations of 0.68, 0.68 and 0.63. But the goal of space manufacturing is tied tightly to them with correlations in the 0.55 to 0.60 range, and space experiments follows closely behind it.

Four other goals achieved a correlation of 0.50 with one in the order, in each case with space manufacturing, and are listed at the bottom of the table. Two are the familiar technology and spin-off goals, and one cites general benefits for industry. The goal of benefiting physics through space research might seem a strange inclusion, especially if we conceptualize this science in terms of nuclear physics or cosmology. But most physicists, as a matter of fact, work in materials science, developing semiconductors or other materials of value to industry. The manufacture of alloys and crystals in weightless conditions presents exactly the range of problems that this more practical conception of physics would address.

Many respondents to S1986A were aware of "developments made in medical research that is conducted in space," and they felt "a medical lab in space may be very beneficial." "Medical research in space just might have some advantages over normal everyday medical research." We can do "better drug and other research in the gravity-free environment of space." "It can help in research designed to discover cures for diseases that can't be done in our atmosphere." Some noted the importance of "medical spin-offs (for AIDS, cancer, etc.)." "For example, technology obtained from the Apollo program helped improve the kidney dialysis machine." Benefits might also come for "medical testing," "health care," "paramedic equipment," "biomedical research." The "scientific gains could improve life spans" and even, perhaps, give us "a cure for the common cold." "They might be able to cure diabetes, which I have, through space tech."

The result of such research might be "purer medicines," "better pharmaceuticals," "development of vaccines," and "fabrication of new drugs." "Certain drugs crucial to advanced medical technology or other chemicals may be produced in a zero-gravity environment at a far faster rate than on Earth." "Zero-G technology could manufacture new and helpful drugs." Therefore, there was much support for "creating a space factory producing new drugs" "in a weightless environment," thus using "the peculiar properties and resources of 'space.'"

There was also much interest in the "effect of weightlessness on diseases." "Studies on the effects of gravity on arthritis could lead to improved treatment." "Some people are more easily rehabilitated in lower gravity." We should exploit "special space conditions that help a diseased person" and develop "health care in space." "Send old people into space; reduced gravitational stress would allow them to live longer."

Parallel to manufacture of medicines in space is orbital production of "inexpensive and pure crystals," "silicon chip production," "super-pure chemicals," "synthetics," "plastics," "metal alloys," "perfect ball bearings," "building materials," and other "products made better in space." Industry should "take advantage of weightlessness and high vacuum for production purposes." "Some chemistry works in space better than on Earth," allowing "new ways of bonding molecules," new "chemical reactions" and "pure synthesis of organic chemicals." "Mixing of alloys in space gives a 100 percent homogeneous mixture due to lack of gravity; this has economic and building importance." "Zero gravity engineering" permits "new industrial techniques, new development of fine parts" and "better machines." "Space manufacturing will provide a means of creating many new high-tech products." With the "growth of heavy industry in space," "material science should advance," and "expanding materials capability affects all American life."

Finally, there are great "opportunities for experimentation in a nongravity environment" "beyond our atmosphere." We should launch "zero-gravity labs where various important experiments can be executed." We could achieve "increased accuracy on scientific experiments, especially chemistry. In the vacuum, there's nothing else to disturb the reaction." And these "experiments made in space can help us solve problems on Earth."

THE COMMERCIAL-ECONOMIC ORDER

The three goals of this order speak of commercial applications, opportunities for business, benefits for industry, and economic benefits. The average correlation linking them is 0.56, but their 0.46 average correlation to spin-off goals shows a substantial affinity with that order. These three items, plus four outside the order, are shown in Table 4.4.

Table 4.4. The Commercial-Economic Order

| | Average Correlation | | 106 | 97 | 35 |
	Intra-class	Extra-class	Commer-cial	Industry Benefits	Economic Benefits
106. Space has great commercial applications and many opportunities for business.	0.40	0.27	1.00	0.65	0.52
97. The space program has great benefits for industry.	0.43	0.27	0.65	1.00	0.51
35. The space program stimulates the economy and has direct economic benefits.	0.35	0.28	0.52	0.51	1.00
Related Goals:					
15. The space program contributes much to our technology.	—	—	0.51	0.56	0.45
33. Technological spin-offs (advancements developed for the space program, then applied to other fields) improve every-day life.	—	—	0.51	0.54	0.45
98. In the weightlessness and vacuum of space, we could manufacture new and better alloys, crystals, chemicals, and machine parts.	—	—	0.49	0.51	0.38
90. The space program provides jobs for thousands of people.	—	—	0.36	0.43	0.50

The first two goals in Table 4.4—opportunities for business and benefits for industry—are very closely connected, as their correlation of 0.65 proves. The item about economic benefits is somewhat less tightly connected, at 0.52 and 0.51, perhaps because it does not mention by name the institutions of capitalist society, business and industry. At the bottom of the table we see that business and industry connect to technology, spin-offs, and manufacturing. Economic benefits connect more strongly to "jobs for thousands of people," aiding the inhabitants of industrial society, not its institutions.

Respondents to S1986A gave clear if simple statements of these goals. "There are great commercial applications," "new industries and new capabilities," and "endless possibilities for business applications" in space. "The commercialization of space seems like a move in the right direction," and we should "explore possible uses of space for commercial reasons." "Opportunities for business would increase greatly if we developed space," and already "it keeps many large American companies in business." "The space program has commercial benefits, such as satellite launching for private interests or other countries," yet "commercial benefits of space research are often overlooked."

Other utterances spoke simply of "industry," "industrial development," "industrial and economic gain," or "industry, thus economy, benefits." As one defined the concept, "Space industry: People who manufacture rockets, etc., make lots of money."

"The very obvious material benefits" are not just "economic benefit for space-related industries" but contribute to "a more productive economy." The space program "boosts GNP," "stimulates the economy," achieving "expansion of economic frontiers" and "increased living standards." True, "it brings business to East-Central Florida," but it also contributes to the "economic well being of the world."

THREE SCIENCE ORDERS

The next three orders are small enough and similar enough in concept to be treated together; their seven goals share Table 4.5.

The first of the three general science goals is one of the stray life-related items I mentioned earlier, which might have joined the three medical items in the Exploitation of Space Conditions order: "We could learn more about life by experimenting with life processes in the different conditions of space and other planets." Its highest correlation is 0.45 with "Space research tests our scientific theories and promises conceptual breakthroughs." As the table shows, it does almost as well with the physics item. Clearly, this item reflects aspects of several other goals, and is not well classified with any particular group, even if our method placed it here. The representative item in the order is the simplest: "The space program contributes to the advancement of science."

Table 4.5. Three Science Orders

| | *Average Correlation* | | | | |
	Intra-class	Extra-class	31 Science	122 Telescope	78 Physics
General Science Order:					
39. We could learn more about life by experimenting with life processes in the different conditions of space and other planets.	0.37	0.25	0.41	0.35	0.43
38. Space research tests our scientific theories and promises conceptual breakthroughs.	0.40	0.24	0.59	0.42	0.49
31. The space program contributes to the advancement of science.	0.40	0.25	1.00	0.40	0.46
Astronomy Order:					
122. An orbiting space telescope could give astronomers a much better view of the stars.	0.39	0.24	0.40	1.00	0.53
58. Space probes increase our knowledge of space, planets, comets, and the entire solar system.	0.37	0.24	0.48	0.57	0.50

Table 4.5. *(continued)*

	Average Correlation				
	Intra-class	Extra-class	31 Science	122 Telescope	78 Physics
Physics Order:					
78. Space research benefits physics—in studies of the nature of matter, for example.	0.43	0.23	0.46	0.53	1.00
100. We could gain a better understanding of the universe as a whole and how it functions.	0.35	0.26	0.39	0.52	0.56

The average correlation among the seven goals of these three orders is 0.47, and we might want to reconceptualize the group as one order divided into three families. The average correlation linking the General Science triad is only slightly higher, 0.48. But the four remaining goals achieve correlations of 0.57 and 0.56 within their pairs. Because we are dealing with a small number of goals, it does not matter much whether we call them families or orders. If conceptualized as an order, the Science group is relatively cohesive, its internal average correlation of 0.47 contrasting with the average 0.37 to the 23 in the first four orders, described earlier.

The open-ended material on which the goals are based is quite solid. Respondents to S1986A saw that "the life sciences benefit tremendously" from the space program. "Experimentation with life in space" and "experimenting with life on different planets" allow us to study "behavior of many life processes in a reference frame other than our own." We can "grow new strains of virus" and thus "expand our own knowledge of life."

"As a sort of test for some of our theories that might otherwise never be tested—from Einsteinian physics to computer technology—the space program has been very valuable." In addition to the "testing of scientific theories," space research may mean "discovery of a new paradigm for our concept of the universe, physics, and reality. A scientific revolution."

A total of 206 utterances spoke generally of the scientific gains from space research, using phrases like *scientific progress, scientific discovery, scientific advancement, breakthroughs in the sciences,* and even *the march of science.* Several stated their own particular creeds of progress. "I think scientific knowledge should be continuously expanded in all directions." "Pure and applied scientific research benefits—both make the space program beneficial." "The value of pure scientific research alone justifies the space program." "Scientific knowledge always yields benefits." "Most of the universe is outside the Earth and must be explored to further scientific knowledge." "It will broaden our scientific knowledge beyond imagination." We should explore onward "to advance our understanding of the chemical, physical, and biological sciences." "To do scientific research. *Not* propaganda, *not* defense, but scientific research in space." "People talk glamor and money since that's what makes the world go around. However, the scientific benefits, such as the discoveries of Skylab, outweigh any monetary challenges, in my mind."

The two methods of doing astronomical research in space, space telescopes and space probes, are closely related in respondents' minds, as the 0.57 between them in Table 4.5 indicates. In addition to one misguided S1986A respondent who hoped for "advances in astrological knowledge," many saw great potential for "astronomical advances," "astrophysical research," to gain "better astronomical knowledge of the system," "to learn of the stars," and to "learn more about the galaxy." "The planned space telescope" will be "sublime bliss for astronomers."

Space exploration must continue "for purposes of discovery, learning more about the universe. Voyager is a good example." "The most important reason is that through this program, we gain information about other planets and space itself, which gives us information about our past, present, and future existence." More probes should be sent, to get a "better picture of Halley's Comet," "to advance man's knowledge of the solar system," thus "acquiring knowledge about things beyond Earth." "A quest for and accumulation of knowledge of the world and solar system is essential to our existence."

The Physics class, like Astronomy a mere pair of goals, is

held together by a powerful 0.56 correlation. In the view of S1986A respondents, space research, such as "study of the nature of matter," is greatly "for the benefit of physics." "An incredible amount of scientific data has been acquired through the space program. Many theories of physics have been confirmed or refuted by data collected from the space program."

"We should continue it to help us understand how the universe functions" "and the way it all started." "The universe is a big place; why not find out a little of what's in it?" "People like to look up and know what's there." We should seek "increasing knowledge about the universe, without which we are worthless." "From a scientific point, it is important to continue the quest into the unknown and to understand the universe."

THE LONG-RANGE BENEFITS ORDER

The final order of goals in the Technical class consists of a pair of optimistic statements about our future in space: "The long-term, ultimate benefits of the space program could eventually be important." "Space could offer many unexpected benefits we cannot now foresee." Table 4.6 documents the strong connection between these two and lists the four other goals that achieved correlations of 0.45 with either of them.

I find it fascinating that this pair is most closely connected to the seven Science goals, with an average correlation of 0.40, and next most closely associated with the Spin-off Technology order at 0.39. Two of the four related goals in Table 4.6 are the advancement of technology and science. Apparently, this is what space progress means to our respondents. Space exploration is an integral part of general scientific and technological development. The inclusion of "valuable, practical information" stresses that respondents feel the future of spaceflight is not a wild dream but a sober source of data and ideas. These three goals are all members of the Technical class. The final item, about "limitless opportunities," has some affinity with the Idealistic class of goals, but on balance seems to belong to the Colonization class, and thus it will be discussed in Chapter 7.

Table 4.6. The Long-Range Benefits Order

	Average Correlation		9	113
	Intra-class	Extra-class	Ultimate Benefits	Unexpected Benefits
9. The long-term, ultimate benefits of the space program could eventually be important.	0.35	0.24	1.00	0.54
113. Space could offer many unexpected benefits we cannot now foresee.	0.36	0.28	0.54	1.00
15. The space program contributes much to our technology.	—	—	0.49	0.43
31. The space program contributes to the advancement of science.	—	—	0.46	0.42
89. Space research provides valuable, practical information.	—	—	0.50	0.43
94. Limitless opportunities could be found in space.	—	—	0.43	0.45

In a way, these two goals are more abstract than the others of their class. They refuse to identify narrow, specific targets for human action. Just five S1986A respondents explicitly valued the "ultimate benefits more than present benefits." One noted that "people tend to be too shortsighted when it comes to the government budget," and another stressed that "the long-term benefits outweigh the short-term."

Twenty-one expressed faith that "unexpected knowledge" or "unforeseen potential benefits" would come from further space development. "Columbus explored an earlier frontier. Could he have listed all the reasons why it was important for him to explore?" "The frontier argument—the fact that we cannot foresee what values it may have but that preparation for the unforeseeable future is essential." "Sometimes we get the answer to things by studying something completely different. Who knows; perhaps a cure for cancer can be found in the dust of Saturn's moon." "Space offers many benefits we haven't even thought of. I'm sure, as time goes on, better and better ways to take advantage of the environment of space will be developed." "We don't know what it can lead to. Possibilities are endless." "Space is an almost completely unexplored frontier. It may hold things which could answer questions about Earth or somehow improve Earth." We can hope for "possible 'surprise' discoveries which could explain unanswered questions (especially Physics—e.g., unity of forces)." "We might get lucky and really learn something." "Who knows!"

THE ENVIRONMENTAL SATELLITE

Orbiting in the vicinity of the Technical class of goals, but associated also with Colonization, is the Environmental Satellite. Its four goals are listed at the top of Table 4.7, and five related goals that achieve correlations of 0.40 with one of its members are listed at the bottom. Solar power stations and new fuels found in space concern the same issue, the terrestrial energy crisis, so the correlation of 0.52 between them is no surprise. But both are connected above 0.40 with an item about controlling pollution, and generation of power in orbit has often been suggested as a way of reducing pollution on Earth. Connected to the pollution item, but not strongly to the energy items, is one about the ultimate environmental degradation, nuclear winter. One could speak of this group as the energy pair with two tagalongs, but perhaps the word *environmental* fits the entire quartet.

In terms of its correlations, the quartet of goals does act like an independent class. The average correlation linking the items contained in it is 0.39, whereas the average correlation to the 121 items that are not members is 0.21. The ratio of intraclass to extraclass correlation is 1.86, and the ratio of their squares is 3.45. The connections to other classes operate through correlations to a few specific members of them.

The five related goals are in two very clear groups. First, we see two Colonization items about exploitation of extraterrestrial minerals and raw materials. Exploitation of new fuels is practically the same concept. And at the bottom of the table we see three medical items from the Technical cluster. Clearly, these pick up the environmental health aspects of the goals in the satellite.

Table 4.7. The Environmental Satellite

	63 Solar Power	72 New Fuels	62 Control Pollution	112 Nuclear Winter
63. Solar power stations in orbit could provide clean, limitless energy to the Earth.	1.00	0.52	0.44	0.25
72. New fuels found in space or the development of fusion power in space could help solve the Earth's energy problem.	0.52	1.00	0.42	0.29
62. From space, we could find new ways to control pollution and clean up our environment.	0.44	0.42	1.00	0.45
112. Through space research we learn the true extent of devastation that nuclear war would bring: nuclear winter.	0.25	0.29	0.45	1.00
Related Goals:				
40. We could find new mineral resources on the moon, Mars, or the asteroids.	0.38	0.55	0.34	0.15
121. We could use raw materials from the moon and planets when natural resources are depleted on Earth.	0.34	0.48	0.31	0.12
79. Medical research performed in space could benefit human health.	0.39	0.40	0.39	0.22
12. New medicines could be manufactured in the zero gravity and vacuum of space.	0.40	0.43	0.36	0.23
99. Some medical problems could be treated more effectively in the weightlessness of space.	0.42	0.42	0.41	0.24

The utterances from S1986A on which these items were based did not explicitly draw connections between them, except in noting the potential cleanliness of solar power. "Solar power stations in space can develop a lot of clean, useful power, greatly decreasing the pollution burden on our atmosphere." "A large power generator in space, transmitting cheap energy to Earth would be absolutely fantastic." "Solar power satellites" "using large panels to collect sunlight" would harness the "all but unlimited energy from the sun" with which "we could solve our energy problem."

Alternatively, we might make "the discovery of a new source of energy," such as "fuels to be found in space"—"a planet of oil, for example"—or we could develop "fusion." "A new energy source/process to replace coal/oil/gas" would be very valuable, as would "discovery of a safe, renewable energy supply."

The space program may also help with "new ways to control pollution" and "ways to preserve our environment," such as "something that will work to help repair the ozone layer." "We need to find ways to clean up our planet," to achieve a "cleaner, better understood environment."

Just two respondents credited the space program with "discovery of the greenhouse effect and nuclear winter" and "learning the true extent of nuclear devastation." The greenhouse effect is demonstrated by the planet Venus. Only a bit closer to the sun than Earth, it is hellishly hot, because like a greenhouse its atmosphere traps solar energy. The nuclear winter thesis asserts that atomic war would fill the air with dust, thus blocking the sun's rays and causing a dangerous drop in both temperatures and food production (Turco, Toon, Ackerman, Pollack and Sagan 1983; Sagan 1983).

THE ORIGINS SATELLITE

If the Environmental satellite orbits between Technical and Colonization, a smaller satellite moves between Technical and Idealistic. Stated at the top of Table 4.8, its two goals advocate discovery of our own origins and the origin of life, thus making this clearly an Origins satellite. Strongly tied to each other, these two are also connected to two goals of similar meaning included at the bottom of the table. The first of these related goals, gaining greater understanding of the world we live in, is a member of the Idealistic class. The other, gaining a better understanding of the universe as a whole and how it functions, is in Technical.

Table 4.8. The Origins Satellite

	43 Learn Our Origin	76 Learn Life's Origin
43. We could discover our own origins, learning about the history of the universe and Earth.	1.00	0.58
76. Through the space program we could learn the origin of life.	0.58	1.00
Related Goals:		
65. We could gain greater understanding of the world we live in.	0.51	0.44
100. We could gain a better understanding of the universe as a whole and how it functions.	0.53	0.47

One could describe all four of these as a satellite, associated with both Idealistic and Technical, but because there is no reason why we have to stick to astronomical or biological metaphors, a better way of describing the quartet is as a bridge. Understanding how the universe *functions* is a job for science. Greater understanding of the world we live in is both more social and potentially more philosophical. It encompasses the humanistic question of what the universe *means*. Some respondents believe that scientific analysis is the best form of understanding, whereas others have social empathy, aesthetic appreciation, and literary sensitivity in mind. Our origins and those of life can be approached from an entirely scientific viewpoint, or they can be construed in religious, philosophical, or even psychological terms. To call the two Origins goals a satellite is merely to state their relative independence of the two classes they hover near; they are drawn toward the two classes by connections to an item more deeply embedded in each.

For the S1986A respondents, the Origins items invoke profound, philosophical questions. "The scientific advancement the space program provides may lead to the fundamental questions: 'Where did we come from?' 'When?' 'How far does space extend.'" "The program can help us find out about where we came from," "to learn of our origins in space and time." "We can discover the origins of our Earth," the "origins of the universe," "to learn more about our own origins and place in the universe." "We have gained some knowledge of the origin of the universe through probes, etc.," and we must continue "learning about the history of the universe and Earth," "learning about the creation of the universe." And, finally, space research may lead to "resolution of the origins problem," "discovery of the origin and mechanism of life."

SERVING PRACTICAL NEEDS

Twenty years ago, Ordway, Adams, and Sharpe wrote *Dividends from Space*, a comprehensive statement of the practical benefits that the space program had then already partly achieved for mankind. Their chapter titles constituted a categorization scheme for these dividends, and the first two primarily covered spin-offs: "New Products for Home and Industry," "Dividends for Health and Medicine." Four chapters covered space applications involving Earth-related data collected in space: "Observing the Earth from Orbit," "The Oceans," "The Land," and "The Atmosphere." Another discussed comsats: "Communications on Earth via Space."

The concluding chapter looked toward future possibilities under the rubric, "Research in Space." Its topics were solar astronomy, probing the moon and planets, cosmic physics and astronomy, materials and chemical processing, biological research, solar power generation, exploitation of extraterrestrial resources. A few of these goals are so distant that they appear in our Colonization class, but almost all topics covered in *Dividends from Space* are included within the Technical class of goals.

One chapter of *Dividends from Space*, however, concerned a topic that has not shown up here: "The Space Systems

Approach to Problems on Earth." Ordway, Adams, and Sharpe contended that one of the most valuable benefits from space research "is a technique for the management of large, complex, and expensive projects that produces the optimum product in the shortest time at best cost" (Ordway et al. 1971, p. 63). They describe this technique only briefly, but basically it involves a "systems approach." Indeed, NASA and other high-technology government agencies adopted highly rationalized systems of planning in the late 1950s and 1960s, but the most famous and influential was PERT (Program Evaluation and Review Technique), developed for the Polaris missile program, not the space program. Although PERT gained great prestige from the success of Polaris, it contributed little or nothing to that success, and "the space systems approach" was in great measure simply the management fad of its period (Sapolsky 1972).

Today, the fad is Japanese management techniques (Ouchi 1981). Business executives undertake expensive investments of time and personal career in risky ventures. Even well-established operations continually face unexplained loss of market and unanticipated new forms of competition. Much of the most popular management literature can be described as inspirational or revivalistic, using apparently logical but often irrelevant theories of management to revive the executive's confidence and reenergize his or her leadership. As Malinowski (1948) observed about the "savages" of the Trobriand Islands, when people undertake risky ventures under uncertain conditions, they tend to turn to magic for help. Although various management techniques may have objective virtues, often they serve as a kind of magic to comfort and encourage managers. Today the management magic is Japanese; yesterday it was space systems.

Like many fads, the temporary mania for space system management techniques left an enduring legacy. Planning techniques such as PERT continue to be used, but they have become an ordinary part of good management procedure, rather than a distinctively space-oriented way of doing things. Boosted by the spectacular success of Apollo, the managerial prestige of the space program had sagged long before the Challenger disaster unfairly dashed its public relations image. When asked to name the benefits of the space program, respondents do not immediately think of a management philosophy or set of planning techniques. Instead, they think first of practical benefits to technology, science, and the economy.

Although each specific technological and economic benefit seems limited, there are so many of them that they constitute one of the major factors working for growth in our economy. Ben Bova (1981) has argued that we must invest in a vigorous program of space development if we are to continue economic and social progress.

The goals of the Technical class have already been substantially achieved. There remains room for improvement and expansion, of course. We can anticipate future spin-offs and future scientific discoveries. The promise of space manufacturing has yet to be fulfilled, and a few other practical benefits of space remain for the future. But some of the most popular goals of the class were first met years ago. Our questionnaire respondents have grown up in a world served by meteorology satellites and orbital communication systems. Their school textbooks taught them facts about the universe discovered through the space program, and their whole lives have been spent in an economic system stimulated by space development.

The goals of the Technical class encourage further investment in space projects with immediate payoff. If our society continues to find most of them attractive and if benefits of this class continue to return to Earth, we will gradually develop an ever-greater presence in low orbit. Satellites designed for communications and Earth observation will continue to proliferate in synchronous orbit. And the scientific gains may justify occasional robot probes into the solar system. But these normal, practical motives are not sufficient to drive mankind out into deep space. The benefits they seek are too limited, too Earthbound, too utilitarian. We must look to a more revolutionary class of goals if we are to find motives of a transcendent nature.

CHAPTER FIVE

IDEALISM AND EMOTIONAL MOTIVES

Rational calculation of short-term economic advantage, as every self-sacrificing parent can testify, is not the only basis of human investment. And many of the most important historical transformations were shaped by nonutilitarian motives (Pareto 1935). The extreme example is capitalism itself, which sociologist Max Weber (1958) suggested was in part the result of Protestant religious ideals. The previous chapter identified a class of goals for spaceflight that combined the expansion of practical knowledge through scientific progress with the near-term exploitation of space for terrestrial economic benefit. Now we will consider a set of goals that stress ideals and emotions, the Idealistic class.

In the following discussion, I will divide the fifty-five goals of the Idealistic class into fully ten orders, and some of them are conceptually quite remote from others. Yet all speak of ideals, and most are strongly connected through the network of correlations. The average intraclass correlation for these fifty-five is 0.36, compared with an average extraclass correlation of 0.22. The ratio of these average correlations is 1.64, and the ratio of their squares is 2.68.

An average intraclass correlation of 0.40 or greater is achieved by seventeen of the fifty-five. In first place, with 0.47, is: "Spaceflight reaffirms faith in man's abilities." Next, three goals are tied with 0.45: "Space exploration is a human struggle, expressing the unconquerable human spirit." "The space program gives a goal and a feeling of long-term purpose for humanity." "The exploration of space lifts morale and instills a sense of hope and optimism." Two others are tied at 0.44: "Spaceflight is a noble endeavor, expressing the hopes and aspirations of humankind." "Investigation of outer space satisfies human curiosity."

Three goals have average intraclass correlations of 0.43: "Space offers new challenges, and civilization would stagnate without challenges." "Space exploration fulfills the human need for adventure." "Space gives people something to dream about." Another trio have 0.42: "Space stimulates the creative, human imagination." "We must broaden our horizons." "Space is the new frontier." And another has 0.41: "Space triumphs give us justified pride in our achievements." "Humans have an innate need to search and discover." "The beauty of space creates a sense of wonder." And a final pair completes our picture of the essence of the class: "We should explore the unknown." "The space program gives us new perspectives on ourselves and our world."

As in the previous chapter, we can identify orders and families of goals by inspecting the highest correlation achieved by each member of the class. Following this method of cluster analysis, we find that thirteen goals comprise the first group, including the four with highest average correlations within their class. Speaking of faith, spirit, purpose, morale, and optimism, they seek to inspire people with the transcendent value of spaceflight.

THE INSPIRATIONAL ORDER

The largest order of idealistic goals brings its 13 goals together in two or three discernable families. At its heart are two goals that have their highest correlations with each other: "Space exploration is a human struggle, expressing the unconquerable human spirit." "Spaceflight reaffirms faith in man's abilities." Most strongly connected to the first of these, and drawing in three other goals is "Spaceflight is a noble endeavor, expressing the hopes and aspirations of humankind." The goals are listed in Table 5.1, which gives average intraclass and extraclass correlations and the associations with these three family-leading items.

Table 5.1. The Inspirational Order (correlations)

	Average Correlation		125	107	81
	Intra-class	Extra-class	Human Spirit	Noble Endeavor	Faith in Abilities
Human Spirit Family:					
125. Space exploration is a human struggle, expressing the unconquerable human spirit.	0.45	0.25	1.00	0.69	0.71
87. The space program represents the best traditions of Western Civilization.	0.35	0.23	0.53	0.48	0.51
124. The space program gives a goal and a feeling of long-term purpose for humanity.	0.45	0.28	0.62	0.59	0.61
71. Space offers new challenges, and civilization would stagnate without challenges.	0.43	0.27	0.58	0.55	0.59
Noble Endeavor Family:					
107. Spaceflight is a noble endeavor, expressing the hopes and aspirations of humankind.	0.44	0.25	0.69	1.00	0.66
115. Progress in space is part of the advancement of mankind.	0.38	0.29	0.56	0.61	0.55
42. The exploration of space lifts morale and instills a sense of hope and optimism.	0.45	0.24	0.61	0.65	0.65
116. The space program allows people to think beyond the triviality of Earth-bound conflicts and concerns.	0.39	0.21	0.50	0.55	0.53
Personal Inspiration Family:					
81. Spaceflight reaffirms faith in man's abilities.	0.47	0.26	0.71	0.66	1.00
32. Astronauts are heroes and role models for young adults.	0.36	0.22	0.49	0.44	0.56
104. Space triumphs give us justified pride in our achievements.	0.41	0.25	0.58	0.59	0.64

Table 5.1. (*continued*)

	Average Correlation		125	107	81
	Intra-class	*Extra-class*	*Human Spirit*	*Noble Endeavor*	*Faith in Abilities*
105. The space program encourages people to make achievements and solve problems.	0.39	0.25	0.50	0.50	0.53
95. The space program inspires young people to study the sciences.	0.36	0.25	0.41	0.39	0.47

The three families, which emerge when one considers only the highest correlations for each goal, are hardly distinct when one looks at all the coefficients. Within each group, the average of the twenty-two correlations is 0.52. Across groups, the average of the fifty-six correlations is 0.50, only insignificantly lower. Yet perhaps they do identify three slightly different shades of meaning in the thirteen closely connected goals.

The order itself is closely connected to the next one we shall consider, which I call Exploration. Item 125, about the unconquerable human spirit, achieves correlations of 0.50 with seven of the eleven goals in that order, compared with ten of the twelve goals that share its own order. The noble endeavor goal achieves 0.50 with four goals in the exploration group, and the item about reaffirming faith in man's abilities does so with five.

The first goal in the order, concerning the unconquerable human spirit, was based on only two utterances from survey S1986A, the minimum for inclusion in the set of goals, yet it plays a significant role in the statistical analysis. One S1986A respondent praised the "psychological humanist ideal of the unconquerable human spirit," and the other felt the space program was a "poignant human struggle against cosmic impenetrability." Connected to this goal is one we discussed in Chapter 3: "The space program represents the best traditions of Western civilization." Here, not its jingoism but its idealism undoubtedly draws it in.

Third in the order, a goal and a feeling of long-term purpose for humanity, was based on utterances that communicate the idea clearly, but do not expand on it in detail: "Space exploration is a necessary commitment to the future." It "gives the country a purpose for being, a goal." It "gives us something to shoot for, be inspired by." Providing a "feeling of long-term purpose for humanity," it "gives us a goal as a nation and as a race of sentient beings."

"The human race needs challenges, or it'll stagnate," and "America needs new challenges." "Human spirit is stagnating. We *need* a challenge of a new area of exploration." "Growth and expansion of human endeavors is what allows our civilization to continue." Space is "a challenge that man's natural curiosity will always wonder about and want to reach," "an intellectual challenge" "to stretch us, to explore new dimensions." "Our exploration of space cannot stagnate. We must always be looking ahead." The space program "keeps people reaching 'for the stars' to do things they might not otherwise," "preventing the world psyche from slipping into complacence," "keeping the human race 'in shape,' preventing lethargy and the deadening of curiosity."

The statement about spaceflight being a noble endeavor was based on two utterances backed up by two others that simply said "idealism." "It is a noble endeavor," "because it expresses our hopes and aspirations as a race."

Five respondents said spaceflight was simply "progress." Others called it *advance, advance of man,* and *the advancement of society.* Another felt we must "explore areas that might somehow accelerate the development of mankind."

Eleven spoke of "morale"—"morale-building," "good for morale." "People can be inspired by" the exploration of space. "It does instill a sense of optimism in people," "the sustenance of hope"—"national hope," "imaginative hope," a "psychological boost of hope to an often hopeless world."

"The space program gives us something to concentrate on besides our national/international problems; it makes us think in larger terms than 'them and us.'" "It gives people an interest in things beyond themselves," "looking beyond the triviality of man's Earth-bound conflicts and concerns," "beyond our daily perception." "There are intrinsic goods in exploration and human expansion that can't be quoted like a wife quotes her shopping list."

"Spaceflight is a monument, in a time when none of the old ones seem sacred any more. It is a source of hope and pride." It "reaffirms faith in man's abilities." We "need such examples of benign technology to ease society's technophobia," "to demonstrate that man can use his energy for peaceful, cooperative purposes."

The space program "gives people heroes." "Astronauts are safe heroes, healthy heroes. Anyone has the potential to be one." They are "role models for young adults."

Space "triumphs" provide "pride in achievement," "the pride we as a species feel in displaying our intelligence by overcoming nature."

The space program "encourages problem solving." "It encourages people to make scientific discoveries," "inspiring achievement in youth."

"The space program can encourage children to become actively involved in the search for answers in a universe filled with questions." "The goal of being an astronaut often encourages children to become scientists," "makes schoolchildren interested in astronomy." Space exploration "can be used as propaganda to encourage the pursuit of education by children, especially in science." "It serves as a spark for interest in the sciences both at the school level and at the national level, as in the case of Sputnik." It is important for "popularization of science and research. A space program captures the public imagination more than Earth-bound lab research."

THE EXPLORATION AND ADVENTURE ORDER

The second Idealistic order consists of eleven items with a somewhat complex structure. In terms of the highest correlations for each goal, one serves to hold the entire group together: "Investigation of outer space satisfies human curiosity." Tied to this one, as a footnote identifying its emotional quality, is the idea that space is psychologically satisfying. Also connected is the proposition that humans have an innate need to search and discover. Tied to it in turn is the assertion that we should explore the unknown.

The word *should* renders this statement imperative, rather than merely descriptive, and a family is headed by it, in which two of the three other items also containing the word *should*. One is part of the *Star Trek* motto: "We should boldly go where no man or woman has gone before." Another states we should go into space simply because it is there, drawing a parallel to the conquest of Mount Everest. The final member of the family says that we should seek knowledge for its own sake.

The goal of satisfying human curiosity has its greatest correlation with the goal of fulfilling the human need for adventure. Tied to adventure is the creative, human imagination. Next in a chain of connections is the gaining of new perspectives on ourselves, and the sequence ends with understanding our place in the universe. Thus from curiosity to understanding is a chain of five goals including adventure, imagination, and fresh perspectives. These eleven goals are listed in Table 5.2 in three families, and the average intrafamily correlation is slightly higher than the average across families, 0.50 compared with 0.44.

It is noteworthy that curiosity is linked to bold exploration and adventure, rather than to the scientific goals considered in the previous chapter. Goal 14 (curiosity) has an average correlation of 0.32 with the other 124 goals, a bit above the grand mean for all 125 of 0.26. Its connection is hardly higher (0.34) with "The space program contributes to the advancement of science." One would think that the research in astronomy or physics done in the space program would be the main way that our curiosity can be satisfied. But the correlations linking curiosity to space telescope and space probes are not significantly greater than this average, 0.36 and 0.33.

The goal of satisfying curiosity correlates above 0.50 with six other goals: "Space is the new frontier." "Space missions are exciting." "Space gives people something to dream about." "The beauty of space creates a sense of wonder." "New experiences and perspectives gained in space inspire art, music, and

Table 5.2. The Exploration Order (correlations)

	Average Correlation		*14*	*77*	*3*
	Intra-class	*Extra-class*	*Satisfies Curiosity*	*We should Explore*	*Need for Adventure*
Curiosity Family:					
14. Investigation of outer space satisfies human curiosity.	0.44	0.23	1.00	0.59	0.66
5. Space is psychologically satisfying.	0.39	0.18	0.59	0.39	0.55
50. Humans have an innate need to search and discover.	0.41	0.25	0.64	0.60	0.55
Exploration Family:					
77. We should explore the unknown.	0.40	0.26	0.59	1.00	0.47
41. We gain knowledge, and it is good to have knowledge for its own sake.	0.33	0.22	0.51	0.55	0.40
120. We should boldly go where no man or woman has gone before.	0.36	0.22	0.50	0.59	0.46
108. We should go into space for the same reason people climb Mt. Everest—because it's there.	0.35	0.19	0.51	0.51	0.49
Adventure Family:					
3. Space exploration fulfills the human need for adventure.	0.43	0.23	0.66	0.47	1.00
1. Space stimulates the creative, human imagination.	0.42	0.21	0.57	0.44	0.58
29. The space program gives us new perspectives on ourselves and our world.	0.40	0.20	0.48	0.36	0.48
18. Space research helps us understand our place in the universe.	0.32	0.20	0.37	0.35	0.33

literature." "We must broaden our horizons." These are all emotional or idealistic goals for the space program.

The material from S1986A on which the items were based gave little hint of this structure. Thirty-eight utterances used the word *curiosity* to describe the motive of space exploration. "Investigation of outer space will satisfy curiosity" "about our environment and the universe beyond our world." "We're fascinated with what is around us and want to know more," "to know how we got here, how things work."

Exactly how the space program is "psychologically satisfying" was not specified by the seven respondents who contributed to this idea. The "psychological value" might be "a feeling of satisfaction among the populace," "the psychological benefit to mankind from space exploration," or even the "satisfaction" of "forging our place in the galaxy."

"It's only natural that we explore space." "Man is an explorer." "It is in the general nature of man to pursue the unknown." "Research and discovery are important—man's purpose for existence!" "I believe man's destiny is to search, explore, and learn more about our cosmos. This obviously implies the necessity of a space program. It's our key to reaching the rest of the universe." "For humans to continue to distinguish themselves from other creatures, we must search for answers." There is an "almost primordial human need to explore and discover new worlds."

"Just as Columbus set out on a voyage that led to the discovery of a new world, I believe it is our duty as human beings to explore the unknown universe that is before us." "It's important to explore even if the ultimate implications of exploration are not immediately apparent." We should do it "for exploration's sake itself," for the "pleasure of discovery." In the words of Star Trek, we should "'explore strange new worlds.'" "It's a sublime and necessary task for us as human beings."

Fully 146 S1986A utterances said "pure knowledge" was a proper goal of the space program: "to learn," "information," "understanding," "knowledge," "knowledge for its own sake," "greater knowledge is itself a good," "the pursuit of truth," "advancement of human knowledge." "The pursuit of knowledge, especially which probes such questions as whether there is intelligent life out there or how the universe began, is an end in itself." "I'm an advocate of the pursuit of knowledge because it gives existence texture. Imagine life without it." There is an "intrinsic aesthetic benefit of knowledge" "beyond 'ars gratia artis.'" Doesn't 'scientia gratia scientiae' mean anything?" "Science for science's sake!" Or does the Latin better translate as 'wisdom for the sake of wisdom?'

Nine respondents quoted the *Star Trek* motto, "to boldly go where no man has gone before," some adding further parts but all endorsing this as a goal for spaceflight. This famous motto identifies transcendent goals for a future space program; its most recent version reads: "Space, the final frontier. These are the voyages of the starship Enterprise, its continuing mission: to explore strange new worlds, to seek out new life and new civilizations, to boldly go where on one has gone before."

Another nine used the "it's there" argument. "We should explore space for the simple reason that is used for people who climb Mt. Everest. Because it's there." It was George Mallory who first explained the desire to conquer the world's highest mountain by saying, "Because it's there." When Sir John Hunt attempted to explain Mallory's meaning, in his book about the first complete climb in 1953, he said Everest symbolized man's struggle to come to terms with the forces of nature, offered the possibility of entering the unknown, and was a problem that had resisted the skill and persistence of other explorers (Hunt 1954). Mallory, himself, cannot interpret his laconic, enigmatic words for us. He vanished on the mountain in 1924.

Space exploration "fulfills the need for adventure." It provides "the satisfaction of adventure, exploration and even conquest." "The High Frontier—the excitement of adventure and discovery." After the moon landings, real space voyages have ceased to capture the public's sense of adventure. Indeed, NASA seemed to go as far as possible in denying that spaceflight was dangerous, requiring high courage. Only when a schoolteacher was killed on the Challenger did the absurdity of this posture become fully evident. If spaceflight actually is dangerous, and adventure is one of its potential attractions, then depicting it as routine is foolish.

In addition to physical adventure, S1986A respondents saw the potential for adventures of the mind in spaceflight: its "poetic uplift" "spurs imagination," "creativity, critical thinking." "It is an outlet for creative minds and geniuses bored with mundane problems and issues." "By expanding our knowledge of the universe, it helps to increase the imagination and therefore the creativity of the individual."

"The space program challenges every individual to learn more about their world and the possibilities of others." "It gives us newer perspectives," "increased consciousness," "expansion of philosophical frontiers," and "a broader conceptual framework within which to think about the world." "Stepping outside and taking a look at ourselves from a new perspective" helps us "grow in wisdom." From space, we "gain a perspective on ourselves as temporal, earthy (in all senses) beings."

We must continue space exploration "to satisfy a natural curiosity about our 'place' in the universe, how we came to be, and what the future may hold for us." We must continue "to learn more about the universe and our place in it, to gain physical and philosophical answers," and to achieve "increased popular awareness of man's relation to universe." "It is a very new area of science which can lead to radical new discoveries about our position in the universe."

THE INTERNATIONAL HARMONY ORDER

The third order of Idealistic goals concerns international cooperation and world peace, and it is outlined in Table 5.3. Two speak directly of *cooperation* and a third uses the cooperative phrase, *helping us learn from each other*. Thus, we can distinguish this triad as the Cooperation family. The remaining four can be described as the Peace family, asserting that the space program can contribute to world peace and unity. For the seven goals of this order, the average intrafamily correlation is 0.47, compared with an average of 0.45 across families, showing that the two families are hardly distinct, and the average correlation linking the seven is 0.46.

Table 5.3. The International Harmony Order (correlations)

	Average Correlation		34	48
	Intra-class	Extra-class	Cooper-ation	Unites the World
Cooperation Family:				
34. Joint space projects between nations improve international cooperation.	0.35	0.26	1.00	0.63
56. Space promotes cooperation between the United States and the Soviet Union, working together for a common goal.	0.30	0.19	0.65	0.59
22. The space program is an educational tool, helping us learn from each other.	0.34	0.22	0.48	0.41
Peace Family:				
48. The common cause of space exploration unites the peoples of the world and could eventually create a world community.	0.36	0.23	0.63	1.00
55. The space program channels government spending away from destructive weapons, and it is better than wasting money on the military.	0.30	0.17	0.43	0.46
102. The space program contributes to world peace.	0.32	0.20	0.48	0.57
36. The exploration of space is an unselfish quest that could benefit all mankind.	0.35	0.23	0.39	0.48

The utterances from S1986A on which this order was based were not numerous, about fifteen for each of the seven goals, but the underlying idea that space could promote peace and international cooperation came through loud and clear. "International space missions could promote a spirit of world peace and cooperation." "International spirit: a cooperative venture might inject hope into future negotiations." "In space we're all on the same footing." "The exploration of space is a peaceful endeavor that all nations can work on together." "A possible result may be the creation of a worldwide space program in which all countries work together." "To have enemies on Earth join together to explore space" may achieve "depressurization of world politics" and even "global nationalism." "It's a project that can conceivably help unite nations and serve as a common goal."

Three respondents mentioned the Apollo-Soyuz project as one that brought the United States and Soviet Union closer together (Ezell and Ezell 1978). "Possible cooperation with Soviets" can aid "world peace through combined U.S.-U.S.S.R. participation." "Like in 2010, it can bring peoples of different ideals together, working for a common goal." "We should continue the space program as a way of uniting U.S. and the Soviet Union in shared scientific concern, not to divide us by making it military." "If done right, common exploration of space may ease tensions" and establish "a neutral arena in which U.S. and U.S.S.R. interests can complement each other, promoting a much more expansive and general peace."

The space program is "an educational tool." It creates "publicity for things like schoolteachers and minorities." It achieves "increased popular awareness of planet ecology." It "promotes human understanding and interaction." It "brings technology and scientific interests to the mainstream population." "It is a learning institution." Through "expansion of educational frontiers" into space, we can "learn from each other."

Space exploration "makes the world seem smaller" and thus "encourages global community," "some sort of nonnational society." In future it can "unify the planet" and "promote peace between nations by linking everyone onto one planet." "Large projects can become a common goal for many people, unifying them and giving them a sense of pride." "Successful space exploration might lead to unity on our sphere. Attempt to parcel out the cosmos would be ludicrous, and no single country could deal with the logistics alone." Spaceflight "promotes the image of mankind as a whole," "fostering a greater 'world view,'" helping "people to learn to work together across cultural lines."

The space program is "nonviolent" and far "better than wasting excess money on defense." It is "best to channel energy and interest away from weapons and destructive things, toward exploration and scientific discovery." Space is "a better way for government to spend money than the military," "a worthy non-destructive expenditure of money and labor," a form of "government money for creativity." "The space program may employ technocrats who would otherwise end up working directly for the military." "It has the same effect on the economy as military spending (i.e., capital is provided for high-tech research), but it is not military." "It's giving science a reason to continue progress, something that otherwise might only happen in wartime." "More money for space, less for nuclear arms!"

Space exploration will "promote world peace" and "prevent nuclear war," achieving "global security" and "making a peaceful world community." "A vigorous program in which many nations participate would encourage peace on Earth, but there is no benefit to a nationalistic program."

"For all mankind's sake, the exploration of space is a deed of humankind for all people." "The space program is not just for the United States It will benefit everyone in the future. It is unselfish and essentially nonpolitical research." "It seems that no one country can claim that 'space' is its own. I heard people express gladness that the Soviet Union was successful in putting some cosmonauts in space weeks after the space shuttle tragedy. And that accident was just a *tragedy,* not an *American* tragedy."

Don E. Kash (1967) identified fourteen justifications for international space cooperation that were discussed around 1960, seeing them as goals that space cooperation might serve, as well as noting that for many people cooperation was a sacred, ultimate goal that needed no excuse. He described them under three headings: innovative goals, conservative goals, and instrumental goals. Some of the 14 resonate with several in our list of 125, whereas others are keyed to themes quite remote from spaceflight.

Kash's seven innovative goals are different ways that collaborative space ventures might accomplish major political changes. The very *dramatic quality* of the missions might stimulate a new awareness of human unity that would break through old habits of conflict. As *a new threshold,* space gave us the opportunity to start afresh, avoiding old mistakes and leaving conflict behind us. Through joint space missions, nations might learn *the cooperative habit* that would transfer to other arenas and eventually remove the causes of the arms race. In an athletic metaphor, the exertions in space might become *the treadmill* on which we could work out our frustrations and that would absorb our aggressive drives. Joint missions with the Russians might provide *a way through the Iron Curtain* that would establish lines of beneficial communication with the secretive Soviet empire. An international space program could be a means of *refurbishing the United Nations,* thus allowing this organization to fulfill its grand aims. Finally, because it occurs above all parts of the Earth, spaceflight is *the concern of all nations* and if not pursued collectively might lead to increased conflict.

Two conservative goals would use international cooperation to further nationalistic, competitive purposes. The *propaganda value* of American space missions is clear, and conducting them in concert with other countries gives them greater influence. Opening our advanced projects to people from other countries serves *our need for scientists* and contributes to the "brain drain" from other lands that has greatly benefitted the United States.

Kash's five instrumental goals hearken back to those discussed in Chapter 4. Harnessing the material resources and human talent of other nations to space development would *expand the boundaries of science* to the benefit of all. The United States can achieve greater *economy* in its space program in cooperation with other nations, both because some of the investment would be on their shoulders and because joint missions may make more efficient use of the world's spacefaring capacity. On a very practical level, the United States needs the cooperation of other states for its worldwide system of *tracking stations* to monitor space missions. Similarly, *meteorology and communications* via satellite are not limited by national boundaries and will benefit from increased cooperation. Finally, as has been the case in aviation, international space cooperation can open *a new market* for the American aerospace industry. This analysis by Kash shows that one can identify goals on many levels and that apparently separate goals may have many mutual implications and causal connections, especially so in the area of international cooperation.

THE EXCITEMENT ORDER

In contrast to the high idealism of the International Harmony order, a group of six goals asserts trivial personal emotions centering on fun and excitement. Listed in Table 5.4, these goals show strong connections, achieving an average intrafamily correlation of 0.55. The average intraclass correlation is 0.30, rather below that for the 0.36 for the class as a whole. But the average extraclass correlation is also low, 0.15 compared with 0.22 for the entire fifty-five members of the class. Thus, these emotional goals definitely belong with the Idealistic class, although they may not express an idea close to its heart.

Table 5.4. The Excitement Order (correlations)

	Average Correlation		52	91 Want to	25
	Intra-class	Extra-class	Fun	Travel	Exciting
52. Space travel is fun.	0.33	0.15	1.00	0.58	0.66
91. I want to travel in space.	0.34	0.18	0.58	1.00	0.55
25. Space missions are exciting.	0.41	0.20	0.66	0.55	1.00
103. I like watching rockets take off and enjoy space probes.	0.29	0.13	0.51	0.48	0.60
13. Vacations and games in space could be entertaining.	0.28	0.17	0.57	0.42	0.53
27. Jokes and cartoons about the space program can be amusing.	0.14	0.05	0.43	0.25	0.38

In Chapter 3 we noted that 3 of these 6 were among the least-popular of the entire set of 125. We called them frivolous and far too unimportant to justify the large expenditures required for a vigorous space program: "Vacations and games in space could be entertaining." "I like watching rockets take off and enjoy space probes." "Jokes and cartoons about the space program can be amusing." However, the fun and excitement generated by space missions should not be dismissed as entirely without social functions because they allow ordinary people to become emotionally involved in an endeavor far from their daily experience. And the desire to travel in space is a vital part of recruitment to the astronaut corps.

The S1986A utterances on which the class was based do not deserve extensive discussion. A few respondents felt spaceflight should continue "for the public enjoyment," "because it's fun"— "lots of fun for adventure-seeking astronauts," "I think the enjoyment derived from exploration (for everyone) is a justification not stressed. I think this is important." "It would simply be fun to travel in space." "I want to go to space—so if we drop the program, I won't have a job." "I want to work for NASA when I grow up."

With "the thrill of the unknown and danger," spaceflight is "the most exciting form of entertainment Americans have." "I enjoy deep space probes!" "I get to watch the rockets take off out my back window!"

Imagine "humans vacationing on another planet." "Vacations on Mars!" "Weekend cruise on the Martian canals." "Venusian steam baths." "Have the Yale-Harvard game in outer space." I'd like to see "volleyball in space." It would be an "entertaining place to experiment with drugs." If these are the amusing jokes we could expect from an expanded space program, clearly it is not worth the effort.

THE AESTHETIC AND NATIONALIST ORDERS

We can next consider two orders that clearly both speak of high ideals, although their concepts seem lightyears apart: Aesthetic and Nationalist. Listed in Table 5.5, four goals belong to each. The average intraorder correlation linking Aesthetic goals is 0.57, and the corresponding figure for Nationalist goals is 0.50. But the average correlation across these two orders is only 0.26. Thus, this pair of equal-sized orders provides an example of the scope of their class and of the efficiency with which correlational analysis makes conceptual distinctions.

According to utterances on which goals of the Aesthetic order were based, "The awe and beauty of space" give us a "sense of wonder." Space is "full of mystery and wonder often lost in this day and age." "The sense of wonder it creates for the individual, and the unity of men from all over the world who work together in exploration, are intangibles the world is sorely lacking."

Table 5.5. The Aesthetic and Nationalist Orders (correlations)

	Average Correlation			
	Intra-class	Extra-class	61 Beauty	67 Pride
Aesthetic Order:				
61. The beauty of space creates a sense of wonder.	0.41	0.18	1.00	0.26
82. There are wonderful sights in space, such as Earth and Saturn, that we can see in pictures.	0.38	0.18	0.64	0.30
80. New experiences and perspectives gained in space inspire art, music, and literature.	0.39	0.21	0.56	0.21
57. Space gives people something to dream about.	0.43	0.20	0.61	0.40
Nationalist Order:				
67. The space program builds national pride.	0.34	0.27	0.26	1.00
26. The space program increases national prestige, producing worldwide respect for America.	0.30	0.22	0.18	0.67
85. The space program generates national unity, encouraging cooperation between numerous sectors of society.	0.37	0.25	0.29	0.58
11. Competition in space is a constructive outlet for nationalistic rivalries that otherwise would take the form of aggression and conflict.	0.27	0.18	0.19	0.41

Space offers "wonderful sights," including "the neat pictures of Earth that we get from astronauts" and the "pictures of Saturn for example." Even in the best Earth-based telescopes, the planets are little more than dim blurs, and prior to space photography we had to rely on the imagination and skill of technical artists, notably Chesley Bonestell, to show us the marvels of other worlds (Bonestell and Ley 1949; Miller and Durrant 1983). Although Bonestell and other artists went to great lengths to achieve accuracy in their art, many space paintings done years ago are hopelessly wrong. The mountains of the moon are far softer in their contours than Bonestell imagined them, the Martian canals do not exist, and the rings of Saturn are vastly more complex than depicted even a decade ago. The marvelous pictures from space probes have transformed the old astronomical art from realism to impressionism, and our fresh appreciation of the diversity of environments in the solar system challenges contemporary space painters.

Space "stimulates artists. Writers and musicians have done some good stuff with outer space." "It keeps journalists employed" and provides "new subjects and views for artistic expression," "inspiration for poetry," and the "expansion of creative impulse to new fields, e.g. science fiction or art dealing more interestingly with space."

Space "gives people something to dream about, returning to ideals." "It is our only dream, in a secularized society, of heaven."

The Nationalist order was based on many clear expressions of patriotic sentiment. *National pride* was the key phrase in twenty-eight utterances, and a similar idea was expressed in forty-seven others. One spoke of "national and species pride," and other phrases were *American pride, pride in country, national self-image, national ego, patriotism, patriotic fervor,* and *nationalistic fervor.* The space program was seen "as a harmless focus for nationalism," as a "focus for healthy patriotism." "It makes Americans proud of America." "It gives the nation a sense of accomplishment." "Progress in the space program engenders national pride by putting our country ahead of others at exploring a new frontier. This helps in developing a sense of community as a nation." January 1986 gave us special evidence that the space program can "contribute to national pride—national grief over Challenger shows America's commitment and involvement."

National pride is the respect a nation has for itself; national prestige is the respect other nations have for it. "The shuttle is first and foremost a prestige and status item for the United States," "a national example to the world." The space program "is a prestige item; that strikes me as a perfectly good justification." It creates "worldwide respect for American technological superiority." This is "great P.R. for the United States," demonstrating the "United States is a leader nation" to "win the admiration of Third-World countries." "When we colonize the planets, we can give them American names."

"Kennedy's goal of reaching the moon was a unifying factor for the nation. A common national goal rallies people around the flag." "Successful missions have a very unifying effect on otherwise disparate groups." "The space program, especially NASA —a civilian organization—encourages cooperation between numerous sectors of society: business, government, the academic community." Thus, it is "a national rallying point," a "source of unity and spirit for the United States" In short, "it promotes national unity."

Space exploration is a "peaceful form of superpower competition, a painless sublimation of conflict," a "constructive channeling of nationalistic rivalries." "The human drama of scientific competition" is a "healthy ground for nationalistic competition," a "substitution of national competitiveness from belligerent to peaceful causes."

THE RESPONSIBILITY AND FRONTIER ORDERS

A second pair of orders, each a triad of goals, bracket the Idealistic class at a different angle. Listed in Table 5.6, the goals of the Responsibility and Frontier orders have substantial average intraorder correlations, 0.47 and 0.53 respectively, and an average correlation between them of 0.30.

At the end of the previous chapter, we noted that the space program was once famous for the allegedly superior techniques of management it had developed. PERT and similar techniques had economy and efficiency as their aim, and neither morality nor technical standards for setting ultimate goals were included in them. Today, some critics of environmental policies believe we must find a radically new ethos of conservation and responsibility. Respondents to S1986A saw a possibility that the global perspective of the Earth seen from space might help achieve this revolution in values.

Table 5.6. The Responsibility and Frontier Orders (correlations)

| | Average Correlation | | 16 | 101 | 20 |
	Intra-class	Extra-class	Realize Fragility	Broaden Horizons	New Frontier
Responsibility Order:					
16. Space travel makes us realize that Earth is a fragile, unique, unified world that deserves more respect and better care.	0.31	0.17	1.00	0.28	0.25
123. In space, we see how small our world is and thus learn humility.	0.34	0.16	0.54	0.38	0.31
84. Living in space could teach us the value of conservation, efficiency, and environmental responsibility.	0.31	0.22	0.47	0.31	0.27
Frontier Order:					
101. We must broaden our horizons.	0.42	0.28	0.28	1.00	0.59
20. Space is the new frontier.	0.42	0.26	0.25	0.59	1.00
66. We have the capability to explore space now, and we must not waste this opportunity.	0.37	0.27	0.28	0.54	0.48

"One thing that going into space can do is give all of us on this planet the understanding of how small and fragile Earth is." "Increased recognition of the variety and general lifelessness of the universe might give us more respect for our own planet." "Maybe it will someday be possible that everyone travel in in the space shuttle and that every person will have seen the globe in its entirety. Then we might treat it with more respect." "In the early days of the moon walks, the space program provided a perspective on Earth as a whole, an interdependent planet without boundaries." "The principle is the whole Earth, as viewed from space with 'no frames, no boundaries.'" "The idea of 'spaceship Earth'" gives us "a cosmic awareness of our relationships as Earthlings," "greater sensitivity to the needs of fragile Earth," and "increased appreciation of the delicacy and uniqueness of Earth."

Astronomer Harlow Shapley (1963) has written about the "intimations of mankind's inconsequentiality" that come from awareness of the true scope of the universe, and respondents to my questionnaires expressed similar ideas. "By exploring the vastness of space, we come to realize how small and insignificant we really are. Maybe this can make us more rational." "Perhaps in coming to understand how minute a part man plays in the universe, we will finally understand we are all men at base, and then perhaps forgo our differences, and there will be peace on Earth." "Perhaps if we had an inkling of the vastness and beauty of the universe, we might be more humble on our planet." From space we "get a true perspective on the Earth's place in the universe—not the center." This "humbling sense of the scope of the universe" may "lessen human egocentricity." "It is impossible to study space and be entirely geocentric in outlook."

"Maybe the greater efficiency required for maintenance of life in space would help us become more efficient at home." "I think it would teach us a little responsibility. A space station would have to be self-contained, requiring us to learn to deal with our waste products, rather than ignore them." We would learn to "increase energy conservation" and create "more efficient living conditions."

If the Responsibility order looks back at Earth with fresh eyes, the Frontier order looks outward for fresh opportunities. If space is the new frontier, to take advantage of its great opportunity we must broaden our horizons. The phrase *New Frontier* was a slogan describing the Kennedy administration's hopes for progress on all its agenda, but when the Apollo project became the preeminent symbol of Kennedy's ambitions, the phrase naturally transferred to space (Logsdon 1970).

"Looking toward other planets, space and procedures" "expands our horizons." "Man among the stars" will "expand the universe from which we each derive meaning." "It extends the medium in which the human race operates," "broadening our horizons to help eradicate petty bigotries."

"Space is the new frontier." "Space is the final frontier." Space offers "opportunity to continue to expand, a sense of a permanent frontier." Space "allows a resurgence of the frontier. First the Western world, then the Western U.S.A., then space, the final frontier." "The desirable effects of frontier upon society include meaningful and lasting results of greater capabilities." "I think the 'frontier' argument is often neglected in favor of arguments about industrial and military applications. But in the end, we develop new industries because people yearn for 'something better,' and this yearning is equivalent to the desire for a 'frontier.'"

"To be able to explore space is an incredible opportunity which we might not always have." "We stand at the threshold of so much cosmologically and scientifically, to turn back now would be a grave mistake." "What would have happened to other pioneers, and inventors, and thinkers, and anyone else who ever dreamed, if they gave up after one failure?" "We have the technology to continue and should do so." "Having the money for it gives us an obligation to continue research that most nations cannot afford to undertake." "It would be a waste of billions in previously spent dollars if we abandoned it now." "We have tasted the fruit of space exploration and cannot discontinue our work." "Progress in space is the logical extension of humanity's social and scientific advancements, which are in grave danger of being quelled if we are afraid to go on."

We have seen the word *frontier* often in this chapter, and many different meanings have been attached to it. In the famous thesis of Frederick Jackson Turner, the frontier was the wild, western margin of the United States that swept in two centuries across the continent. As Americans exploit the open lands of the frontier, "a new environment is suddenly entered, freedom of opportunity is opened, the cake of custom is broken, and new activities, new lines of growth, new institutions and new ideals, are brought into existence" (Turner 1920, p. 205). For Turner, the most important result was the promotion of democracy. The social chaos and isolation of the new lands prevented tyranny and weakened traditions, so that individual liberty could reign.

But this may not be the pattern for space colonization, if the vast investments necessary for such endeavors require official sponsorship by the home government, if the dangers and scarcities of the new worlds demand collectivization, and if the degree of control by persons in authority remains as high as it has been in space missions to date. Thus, space exploration may produce all the cultural and scientific fruits of the frontiers of the past without stimulating individual liberty. Only after distant colonies are economically autonomous may they become politically free, and even then the extreme requirements for high-technology manufactured goods may force each new society to be far more centralized than the far flung settlements of the primitive American frontier described by Turner.

THE EMPLOYMENT AND INSIGHT ORDERS

Two pairs of goals complete the Idealistic class, both connected to goals in the Technical class and thus acting as bridges to it. It may be arbitrary to include them in the Idealistic class rather than calling them satellites, but in so doing I emphasize the humanistic and philosophical aspects that they share with much of their class. Listed in Table 5.7, they concern the social benefit of jobs and careers on Earth, and the self-knowledge and understanding that may come from a vigorous space program.

The average intraclass and extraclass correlations are not greatly different, casting doubt on these orders' membership in the Idealistic class. But these four goals achieved correlations of 0.40 or greater with several Idealistic goals. Both Employment goals did so with "Space triumphs give us justified pride in our achievements." Perhaps many respondents think of jobs in terms of achievements, as opportunities to strive for valued goals. Employing the talents of scientists and engineers correlated 0.43 with "Progress in space is part of the advancement of mankind," and jobs for thousands of people, a measure of social progress, achieves 0.39. Gaining knowledge about ourselves achieved correlations over 0.40 with 23 Idealistic goals, and its partner in the Insight order did so with 8.

Table 5.7. The Employment and Insight Orders (correlations)

	Average Correlation		90	65
	Intra-class	Extra-class	Jobs	Understand World
Employment Order:				
90. The space program provides jobs for thousands of people.	0.30	0.26	1.00	0.19
86. The space program employs many engineers and scientists who otherwise would not be able to utilize their talents.	0.33	0.23	0.54	0.26
Insight Order:				
65. We could gain greater understanding of the world we live in.	0.29	0.25	0.19	1.00
93. We could gain knowledge about ourselves.	0.37	0.25	0.30	0.52
Related Goals:				
35. The space program stimulates the economy and has direct economic benefits.	—	—	0.50	0.23
100. We could gain a better understanding of the universe as a whole and how it functions.	—	—	0.25	0.53

The two goals of the Employment order naturally connect to parts of the Commercial-Economic order of the Technical class, represented at the bottom of Table 5.7 by the item, "The space program stimulates the economy and has direct economic benefits." But jobs for thousands of people, including many engineers and scientists, have a social dimension as well as an economic one. Some respondents undoubtedly think of the Commercial-Economic goals as accruing primarily to large corporations, and many sociologists believe corporations and ordinary people stand primarily in an adversarial relation to each other. Thus, benefits from space that may be described in the largest sense as economic may have two different spheres of action. Commercial profits benefit corporations, and jobs benefit people. Whether this crude and debatable conceptualization shaped responses to our questions, a distinction can be made between Commercial-Economic and Employment benefits, and the latter may properly belong with the other social benefits, such as international harmony, of the Idealistic class.

S1986A respondents said the "space program provides employment for" "hundreds of thousands of workers." Various phrases suggested the location of the job or characteristics of the worker: "on Earth," "in Texas, Florida," "non-military," "for Americans," "for common people." "My friend Tim is employed by NASA. Without them he couldn't pay tuition." "My roommate's new job is writing software for NASA." "Keep my father employed."

"The space program employs many specialists who without NASA would not be able to utilize their talents as effectively." It provides "employment for MIT graduates and others affiliated with engineering and production," "scientists and researchers." It "prevents massive unemployment of scientists and astronomers" and "keeps physicists off the dole." This gives an "outlet for energies and brain-power which would otherwise go to waste" and will "help maintain a core of scientists and technicians that contribute to our country's technological leadership."

As noted in the previous chapter, the scientific data derived from space exploration constitute improvement in our technical understanding of the universe, but respondents probably have a second meaning for the word *understanding* that is more philosophical. The physical sciences state mathematical laws and explanations for natural phenomena, but humans often seek a more intuitive, even poetic kind of understanding. Thus, there exist very different modes of understanding and knowledge.

To know means either to have factual information about something or to be personally acquainted with someone. Some European languages explicitly make this distinction by using two words where English has one—*wissen* and *kennen* in German and *savoir* and *connaître* in French. Sociologists often use the German word *Verstehen* for the kind of sympathetic understanding that one person can have of another's motives and actions. Perhaps when many respondents speak of *understand-*

ing the world or *knowledge* about ourselves, they mean something similar. They seek to find the human meaning—perhaps even the personal meaning for themselves as individuals—of the phenomena. The product of science is abstract, mathematical formulas and quantitative parameters, whereas the product of *Verstehen* is insight of a kind comparable to that gained from great art or literature.

Some of the utterances on which the first Insight goal is based sound closer to scientific explanation than to philosophical *Verstehen*, however. Perhaps the summary statements extracted from the utterances leave the precise meaning open to the respondent's interpretation, and some placed humanistic rather than scientific constructions on them. The "space program helps us to understand our own biosphere," "to gain a more accurate and more complete understanding of the functionings of our planet," to " gain insight into puzzling processes on Earth," "to find the truth behind the 'birth' of Earth." "Experiments in space can teach us more about our planet," perhaps uncovering "evidence of how the Earth was formed." "I support any effort to learn more about the world we live in."

"The more we understand our universe, the better we understand ourselves." Space exploration "gives a better understanding of the human race," and "it could help us understand ourselves and society's problems better." "By direct research with humans forced to endure isolation, we learn more about the human mind under extreme stress." Space research can "test the limits of human abilities."

In the previous chapter, we noted another bridge between the Technical and Idealistic classes, a pair of goals that also concern the advancement of human knowledge: "We could discover our own origins, learning about the history of the universe and Earth." "Through the space program we could learn the origin of life." The first Insight goal achieved solid correlations with these two, 0.51 and 0.44 respectively. Thus, a different analysis could easily group all four goals into an Origins-Insight class, somewhat independent from Technical and Idealistic, acting as a bridge or satellite between these two larger classes.

THE STRUCTURE OF SPACE IDEALISM

The core of the Idealistic class of goals for the space program consists of four orders achieving average correlations of 0.42 or greater with each other: Inspirational, Exploration, Aesthetic, and Frontier. The thirty-one goals of these four orders include all seventeen with average intraclass correlations of 0.40 or greater. The strongest associations are between Exploration and Frontier (0.46), Inspiration and Frontier (0.46), and Exploration and Aesthetic (0.45).

Three other orders achieve 0.40 average correlations with one or another of these four. Excitement and Responsibility are both tied to Aesthetic, and Nationalist is tied to Inspirational. As noted earlier, the Employment and Insight orders are somewhat separate from the class. The goals of the International Harmony group do not link closely to particular other orders, but their average correlation with the class is strong enough (0.33) to mark them as solid members.

Several of the Idealistic goals assert that space travel gives a new perspective to the astronauts who look back at Earth from afar and to those Earth-bound enthusiasts who participate vicariously in voyages beyond our world. From the viewpoint of space, we see ourselves, our nations, and our planet in a new light. In a recent book, Frank White (1987) reports that astronauts commonly experience "the overview effect," a radical shift in consciousness achieved by seeing the Earth as a unity and from outside the traditional limits of human experience. He documents this thesis with material from a number of interviews, but unfortunately his data collection and theoretical analysis were not conducted in a manner that social scientists would consider systematic. Furthermore, although White considers "consciousness" to be the essential ingredient of any culture, he does not draw upon any of the standard literature on this conceptually slippery topic. Yet, his hypothesis that from the new world-view offered by space exploration will come a series of new civilizations is a stimulating expression of the basic faith of the Idealistic class.

The poet Archibald MacLeish (1978) wrote these lines concerning the new picture of Earth brought back from the Moon by the crew of Apollo 8: "To see the earth as we now see it, small and blue and beautiful in that eternal silence where it floats, is to see ourselves as riders on the earth together, brothers on that bright loveliness in the unending night—brothers who *see* now they are truly brothers." Thus, one lesson that can be drawn from the new consciousness afforded by spaceflight is that we must be better stewards and tenants of our own planet, finding ways to cooperate, to grow in spirit, and to preserve life on our home globe.

In familiar lines, Tennyson (1910) reminds us that our destiny is more than just a harmonious home. "Come my friends, 'tis not too late to seek a newer world." The words are spoken by Ulysses, after his great voyage, after the grandest of adventures, after reclaiming his throne from enemies, when all would expect him to relax slowly into a peaceful retirement within the confines of his palace. Instead, he calls for a ship and a crew to embark, once again, "to follow knowledge like a sinking star, beyond the utmost bounds of human thought." The Idealistic class of goals in space seeks both a happier and more humane Earth, and a continuation without end of the noble odyssey of exploration. These are revolutionary goals.

CHAPTER SIX

MILITARY APPLICATIONS OF SPACE

The Technical goals of spaceflight are highly popular, in part because they are immediate and undeniable, whereas the Idealistic goals are sufficiently vague and uncertain that they draw far less enthusiastic support. But one category of goals is both undeniable and debatable, that concerning Military applications of space, which draws both enthusiasm and condemnation. The technical practicality of spy satellites, for example, is not in doubt. But my respondents have reached no consensus about the benefit or harm they offer. Six of the 125 space goals in our list can be described as Military, and their relatively low average popularity is testimony to the sharp disagreements about the defense establishment.

The Military class of goals is highly cohesive. The average correlation linking the six members is 0.55, whereas the average extraclass correlation is only 0.16. The ratio of these numbers is 3.44, and the ratio of their squares is 11.82. In some respects, the class behaves like an order within a larger class. Its items are especially close in meaning to each other, and clear divisions inside the group are hard to discern. But there is no larger class of goals that includes them, and the public debate over the military uses of space is so intense that we must consider them in a chapter of their own.

Again, we can define the class by examining which goals have the highest average correlation with other members of the group. In this case, all six achieve average correlations above our habitual cut-off of 0.40—all indeed above 0.45. In first place, with 0.61, is: "The space program contributes to our defense." At essentially the same level, 0.60, is "There are great military applications of space." Whether we call it the Military class of goals, as I have done, or use the softer word, defense, we shall not go far wrong in describing it.

Table 6.1 shows the average correlations for each military goal, and connections to the three of them that state the most distinguishable concepts: SDI, keeping up with the U.S.S.R., and reconnaissance satellites.

Table 6.1. Goals of the Military Class (correlations)

	Average Correlation			21	30
	Intra- class	*Extra- class*	*96 SDI*	*Keep up with USSR*	*Recon- naissance*
109. The space program contributes to our defense.	0.61	0.14	0.67	0.62	0.52
37. There are great military applications of space.	0.60	0.12	0.70	0.60	0.44
96. A space-based anti-missile system, part of the Strategic Defense Initiative, could reduce the danger of war and nuclear annihilation.	0.56	0.12	1.00	0.50	0.50
21. The United States must develop space technology vigorously to keep up with the Soviet Union and other countries.	0.57	0.19	0.50	1.00	0.48
2. If America doesn't advance into space, some other country will.	0.49	0.20	0.41	0.67	0.34
30. Reconnaissance satellites help prevent war and nuclear attack.	0.46	0.17	0.50	0.48	1.00

The Harvard students who responded to the Spring 1986 survey seldom gave military applications or international competition as a justification for the space program, and when they did they failed to say very much about it. Typically, their responses were terse, often the single word *defense*. Others cited "national defense," "civil defense," "defense applications," "defense systems," "defense spin-offs," "technology spin-offs for defense," "national security," and "U.S. protection." Only one specifically cited defense from the Soviet Union, writing, "for *defense*. Watch Russia and destroy their weapons if they launch." The most eloquent proponent of space defense said, "The outer atmosphere serves as an important strategic outpost in the defense of the United States."

As mentioned earlier, I placed defense utterances in a different category from military ones, thinking that their evaluative connotations might be distinctly different, but the respondents to the Autumn 1986 survey awarded my two items the highest correlation between space goals, indicating that they do not generally distinguish the concepts. The utterances that wound up in the military category were only slightly more complex than those in defense. Six respondents simply wrote "military," and several others placed this word in brief phrases: "military activities," "military advantage," "military advances," "military applications," "military security," "military research," "military strength," and even "military defense."

Also named were "military spin-offs." The space program's "research results in significant discoveries for the military," and "there are many things we can learn about other areas through building a space program (i.e., military)." Others saw space as a "strategic base," for "building military power" and "implementation of military plans," perhaps "to help wage war more effectively." "We could move military conflict into space," and "space war is less destructive than nuclear war."

Only fourteen mentioned something about reconnaissance satellites, one referring specifically to "reconnaissance defense." Six used the word *intelligence*, either alone or in the phrases *military intelligence* and *intelligence gathering*. Three referred to "spy satellites," and two mentioned "satellite surveillance of the U.S.S.R." A couple stated the idea in general terms, one citing "photography for military reasons," and the other suggesting "the early detection of war on a major scale and the possibility of preventing it." Thus, our respondents did not express complex ideas about the topic or show much awareness of the accomplishments of reconnaissance satellites in recent years.

Few respondents mentioned the SDI program, but those who did had no difficulty finding a name for it. They were about equally split between those who called it SDI, Strategic Defense Initiative, or Star Wars. One saw it as "a laser net versus nuclear bombs," whereas another hoped for "a perfect defense shield," and still another wrote of "protection of U.S. citizens in case of nuclear attacks." One imagined "reduction of the threat of a nuclear exchange, by means of a space-based antimissile system," and another said "an antinuclear defense system could eliminate the threat of nuclear war."

Respondents foresaw protection against more than just missiles, perhaps forgetting that the only atom bombs actually used in war were delivered by aircraft, including an "end of nuclear weapons due to defensive systems." SDI could "eliminate usefulness of nuclear weapons, which would then become obsolete and could be dismantled." "If something could be developed that would render nuclear weapons harmless, something that could prevent or inactivate them," it would "free us from the danger of nuclear annihilation." One wrote of "antinuclear defense weapons that shifted military focus back to more traditional weapons"; another also imagined "the use of defense systems to defend against missiles carrying conventional weaponry."

Only one respondent considered SDI in terms of deterrence, describing it as "prevention rather than having to deal with the aftermath of nuclear war. Deterrent—increased safety when completed." Although several respondents saw some value in SDI "to defend against nuclear attack," one interpreted *defense* to mean "ability to win the war in space." Only one even vaguely commented on the consequences of atomic war, saying a successful SDI "eliminates consequences of nuclear winter, i.e., loss of life on earth."

The goal of keeping up with other countries and the concern that other countries will advance into space if America does not drew few utterances from the S1986A respondents, but some of them expressed themselves clearly. "Whatever country gets fluently into space will be the dominant power of the next century." We need "to keep our dominant, world leader position in space," "in the national interest," to achieve "political superiority" or "to help America remain in the forefront of space technology."

Fourteen mentioned competition with "Russians," "Soviets," "U.S.S.R.," "superpower competition," or "Commies." Whereas one was worried that "the Russians might get ahead," another already felt "the Russians are ahead." "We must maintain a more advanced space technology than that of the Russians," perhaps "to get (keep) the upper hand," to gain a "psychological edge over Russians," or "to remain technologically competitive with chief political rivals." One hoped that "we can use our technological advantage to gain a strategic advantage." Another wrote of "the cold-war imperative that resources discovered first to the second-world might not be available to the United States." For one, space developments "represent this country's commitment to stay ahead of malevolent, ill-wishing countries such as the U.S.S.R. in technological advancement, because they would put the knowledge to harmful use." Finally, five believe spaceflight is bound to develop. "If America doesn't do it, some other country will." "If we don't do it, somebody will, so why not us?" Besides, "We don't want to look like dummies."

DEFENSE, NATIONALISM, AND INTERNATIONALISM

Two orders of goals in the Idealistic class concern issues of nationalism and internationalism, and thus should be associated with the Military class, which stresses international competition. The International Harmony order contains seven goals, divided into two families: Cooperation and Peace. Typical goals representing these two families both stress world unity: "Joint space projects between nations improve international cooperation." "The common cause of space exploration unites the peoples of the world and could eventually create a world community." The goal that speaks most directly about peace does not rule out the military doctrine of peace through strength: "The space program contributes to world peace." But its 0.41 correlation to an item in its family about diverting funding from destructive weapons suggests it expresses pacifist sentiments. A typical member of the Nationalist order is "The space program builds national pride."

The average correlation linking goals in the Military class with those in the International Harmony order of the Idealistic class is 0.05, and the same average correlation is achieved with both of International Harmony's families. This is indistinguishable from dead zero. But we should remember that a small but significant positive correlation is likely to connect all 125 goals because each is an expression of support for the space program. So this 0.05 may really reflect a modest negative correlation between Military and International Harmony, pushed insignificantly on the positive side by the common theme of enthusiasm for spaceflight.

One of the Peace family of space goals specifically separates the space program from military aims: "The space program channels government spending away from destructive weapons, and it is better than wasting money on the military." This antimilitary item has an average correlation of -0.06 with the six Military goals, but, again, this may really represent a substantial negative association drawn toward zero by the positive correlation running through all the goals. One of the goals in the Environmental satellite also appears antagonistic to the Military goals: "Through space research we learn the true extent of devastation that nuclear war would bring: nuclear winter." Its insignificant 0.05 correlation with the group may also really indicate a negative association similarly shifted in the positive direction.

The average correlation linking the Military class with the Nationalist order of the Idealistic class is 0.35, substantially on the positive side. Indeed, one might want to remove Nationalism from Idealism and combine it with Military in a new two-order class. But Nationalism has a bigger average correlation with the Inspirational order, 0.41, which in turn is firmly embedded in the Idealistic class.

Table 6.2 shows the correlations connecting four representative Military goals with the four of the Nationalist order. The first two rows contain big coefficients, proving that pride and prestige are the aspects of Nationalism that link most closely with the Military.

Table 6.2. Military and Nationalist Goals (correlations)

	109 Defense	96 SDI	21 Keep up with USSR	30 Recon- naissance
67. The space program builds national pride.	0.42	0.35	0.50	0.36
26. The space program increases national prestige, producing worldwide respect for America.	0.45	0.41	0.58	0.38
85. The space program generates national unity, encouraging cooperation between numerous sectors of society.	0.27	0.24	0.34	0.28
11. Competition in space is a constructive outlet for nationalistic rivalries that otherwise would take the form of aggression and conflict.	0.21	0.19	0.30	0.27

An early study of the American space program summarized its rationale in two words, *pride* and *power* (Van Dyke, 1964). Prestige is one component of power, and after the Soviets launched Sputnik I in 1957, the international status of the superpowers rested greatly on their achievements in space. And the shock waves from Sputnik I wrenched the U.S. from its previously unquestioned leadership in science and technology.

Harold Leland Goodwin (1965) analyzed extensive poll data to show how space had become an important instrument of propaganda in political warfare. A month after Sputnik I, polls by the U.S. Information Agency asked "All things considered, do you think the U.S. or Russia is ahead in scientific development at the present time?" A majority of Britons, 58 percent, felt Russia was ahead. Pluralities in France, Italy, and Norway felt the same way, and only in Germany did more respondents give the United States the advantage in science. On average across the five nations, 24 percent said the United States was ahead, compared with 44 percent for Russia (Goodwin 1965, p. 45). Note that the question concerned "scientific development," not "space development." A defeat in space had damaged the honor of American science and shaken national prestige (cf. Almond 1960; Michael 1960).

THE STRATEGIC DEFENSE INITIATIVE

If the launching of Sputnik I was a propaganda attack and Apollo the ultimate counterattack, thirty years after the first Earth satellite the United States had taken the offensive in the space propaganda war through a highly publicized and controversial military program, the Strategic Defense Initiative. More popularly named *Star Wars* after the movie filled with outer space battles, the program promised to develop a defense against nuclear missiles, greatly based in space.

The idea was not new. Thirty years ago, Austrian rocket pioneer Eugen Sänger (1958) urged the development of space-based energy rays to blast nuclear missiles, calling spaceflight "a technical way of overcoming war." I am sure I was not the first to suggest that peace and space development might both be best served if such a system were created under international control, to end the capacity of any nation to launch an attack through space (Bainbridge 1976, p. 241–243).

Support for SDI varies greatly across groups, as a number of polls have shown. A survey of 1,525 American adults, conducted at the beginning of 1985, found men more supportive of SDI than women (*New York Times* 1985). Majorities of both sexes thought the system would work, 68 percent of the men and 56 percent of the women, but women were much more likely to feel it would hurt international relations rather than help them. Of the men, 54 percent said that developing SDI would "make negotiating with the Soviet Union easier," compared with 42 percent of the women. Of the women, 59 percent said "it would make the arms race more dangerous," compared with 48 percent of the men.

In February 1986, the Union of Concerned Scientists, a group that cannot be described as favorable toward SDI, polled 549 physicists, finding that 54 percent felt it was "a step in the wrong direction for America's national security policy," with 29 percent calling it a step in the right direction (*New York Times* 1986). However, a majority of 77 percent felt basic laboratory research on missile defence should be continued.

At the end of 1986, coordinated polls showed that Americans were far more enthusiastic than citizens of two allied nations (*The Economist* 1986). The question was "Do you support President Reagan's anti-missile defence project, often referred to as the 'Star Wars' project?" Although 50 percent of 1,800 Americans polled were in support, the level was only 35 percent of 1,096 Britons and 14 percent of 500 West Germans. One factor, of course, is general approval of Reagan, which ran 50 percent, 26 percent, and 22 percent in the three groups. Another may be sheer nationalism; although SDI aims to protect all the West, it would be primarily American.

We cannot expect exactly the same range of opinions about SDI from our Harvard students as a random sample of the nation would provide, but few groups will have as much influence on the future of our society than graduates of this university. Florence Skelly (1986) compared Harvard alumni with Stanford alumni and the general public through polls done by her organization, Yankelovich, Skelly, and White. The roughly 3,600 alumni respondents were not only far better educated than the average, but far more prosperous. Median household income was $72,000, three times the national average, and one-fifth were millionaires. A quarter were business executives, with between 10 and 20 percent each in government service, education, law, medicine, and science.

One item in the survey asked whether the Star Wars program will increase the risk of nuclear war, decrease this risk, or have no impact upon it. Among the general public, 20 percent felt SDI would increase the risk, compared with 31 percent of Harvard alumni and 32 percent of Stanford alumni. The percentages believing SDI would decrease the risk of nuclear war were closer, 36 percent, 33 percent, and 32 percent. The difference in those favoring Star Wars is made up by the fact that the Harvard and Stanford alumni were less likely to say it would have no impact, which is probably the response chosen by people who had not thought about the issue and had no opinion.

During the time of my surveys, the standard opinion among faculty at Harvard was that SDI was both infeasible and destabilizing. I did not administer questionnaires to the professors, but this was the consensus of rumors traveling around campus. I repeatedly heard that the top Harvard scientists privately said the scheme would not work, and every Harvardian felt himself competent to judge the political implications. Public speeches, by both locals and visitors such as Carl Sagan, weighed against Star Wars.

The issue was not beyond debate. How, some asked, could a technically impossible defense system destabilize international relations? Why did the Soviets oppose the American SDI program if it was a mere waste of U.S. tax dollars? Assuming the U.S. could make a halfway credible system but the Russians could not, was SDI really designed to demote the Soviet Union from superpower status, defeating an enemy rather than making peace?

Some suspected that SDI was actually part of strategic competition with Japan, not Russia. Even a failed system would demand so many developments in so many branches of technology that it would convey a great economic advantage to the nation that attempts it. Furthermore, under the cloak of military security, these technological developments could be denied to the Japanese until American corporations were thoroughly familiar with them and ready to exploit them in the international consumer market. So went one of the more exotic lines of argument.

One question in S1986A sought students' opinions about SDI, without mentioning it by name: "Recently, there has been much talk about building a system of space satellites to defend against nuclear attack. Do you think research on this idea should continue, or should research stop?" There were three responses to choose from: research should continue, research should stop, or no opinion. A space for written comments gave people a chance to express themselves more fully on this complex issue.

Because this book is about the aspects of American culture that encourage development of spaceflight, and this chapter concerns the thinking behind the Military class of space goals, we shall not examine the arguments against SDI more closely than we have already done. Proponents felt "We should have the best defense possible." "The nation owes to its people every type of security," and "defense is an important service which the space program can provide." "A shield is possible," and "SDI is the only way for the U.S. to achieve true security." "Who can object to a purely defensive weapon?" "The fact that this is a defensive system is appealing, and perhaps it will be a turnaround to the arms race."

A few were very optimistic and felt such a defensive system "would make N-weapons obsolete." "I think that a completed Star Wars will eliminate the need for nuclear arms." If the system is possible, "all nations would eventually be freed of the fear of nuclear war." Or if they cannot be entirely prevented, "wars will be fought in space" where they cannot do so much harm. Even if the system could not provide complete security, it would be valuable "as a deterrent to the possibility of nuclear war." In part, it could achieve this by preserving current means of deterrence, "not with the idea that it will make nuclear weapons obsolete, only with the idea of making our weapons less vulnerable to attack."

Proponents praised the system as "a noble alternative to weapons buildup." "What a wonderful, nonviolent form of security!" "Pursuing a defensive system against attack is better than threatening the other side with destruction." "*Defense* against nuclear weapons is an essential alternative to the nuclear balance of terror." "Mutually assured destruction is not a system that will last," and a system of defense satellites might "protect mankind from MAD." "At present we have a mutual suicide pact with the Soviet Union. I think this policy (mutually assured destruction) lives up to the acronym and should be changed if possible."

Some proponents of continued research felt "the system can

be used as a bargaining chip at the strategic arms limitation talks." SDI "may turn out to be infeasible but brings U.S.S.R. to the table to talk seriously about nuclear arms, not just with rhetoric." "Research should continue. It lets the Soviets know we're serious." "SDI provides leverage at the bargaining table," "will make arms control more equitable," and "should work towards nuclear disarmament."

Others were convinced "research should go on as long as we know the Soviets are doing the same." "Unfortunately, we have to stay one step ahead of Russia." "If we don't do it, the Russians will," and "skeptics are ignorant of Soviet policy." "The Soviets are already in space," and "research should continue to prevent potential Soviet techo-breakout." Also, "research is necessary to prevent a Soviet advantage if the ABM treaty breaks down." "Without this, we will suffer the dangers of a lack of strong deterrence—Russian global dominance." "Failure to utilize potential advances in technology could lead to a dangerously vulnerable position." "Defense satellites are a necessary safeguard, as verbal/written agreements are undependable." "Peace talks haven't really done anything to limit nuclear weapons, and we can really only trust ourselves."

Some respondents could be described as weak proponents, in favor of exploring the feasibility but not confident of success. "The event that a *foolproof* SDI system will be produced is unlikely, but one never knows." "Research should continue on determining the *possibilities* of such an investment." "It is important to continue to research. The decision whether or not to use the technology can be made later."

Others felt it would be "unrealistic to stop research—this is unnatural checking of human progress." "Whether or not SDI is workable, research is necessary." "You can't stop scientific progress. We may as well learn what we can do." "I don't think we can reverse gains in knowledge. We can no more stop SDI than we can dismantle all nuclear weapons. It is the knowledge, not the physical presence, of weapons that is important." "You cannot stop curiosity and inquiry in science or elsewhere. Therefore, to cut off funding for research of such an important concept is both naive and dangerous." "Knowledge does not disappear with suppression." "We should never close doors before they are even touched."

In addition, "there are bound to be spin-off benefits." Satellite defense research "bolsters the economy," "provides jobs," "produces usable technology" and "yields valuable scientific discoveries." "Even if the system never becomes operative, the benefits derived from the research are huge." "It provides technology for radar, lasers, guidance systems," and already "there have been many beneficial, nonmilitary advances due to SDI research. Example: laser-treatment of stack effluents to reduce particulates in factory emissions." "Many important inventions came from defense research." "Throughout history, scientific advances have come from military need (or perceived need). In retrospect, we remember only the advance." "Weapons system research has historically led to advancement of technology and solving of social concerns." "Research by itself cannot do harm. Indeed, many of the greatest and most useful scientific advances of the last century were made under the auspices of war research. What is done with the research is another matter, one that should be decided separately."

Proponents anticipated some of the opponents' objections. "SDI's 'impossibility' is declared by politicized scientists. Nothing is totally impossible." "The SDI program does not have to produce a fail-safe system to defend against attack. It merely has to produce a system which is credible." "It is by no means an escalation of the arms race, as many would argue." "I don't believe defensive systems in space will bring space-to-Earth weapons." The program is "hugely important unless even the research side will dramatically increase superpower tensions (which it won't in the near future)." "In no way would such a system be destabilizing, unless the Soviets were allowed to strengthen their lead in this area, and monopolize the ability to defend against attack."

"Ideally, research for defense mechanisms would not be needed, but let's be practical." "Deterrence is at best an interim strategy, and options should be explored." The satellite system "won't be a complete defense, but it can work as a safety device against mistakes" such as "small nuclear accidents (i.e., 1 accidentally launched missile)." "I think this would protect us if countries other than the Soviet Union obtained nuclear weapons," and "any research to stop nuclear proliferation is important."

The Military space goal in S1986B that speaks of SDI says it "could reduce the danger of war and nuclear annihilation." And this goal was rated rather low; its average score on the 0 to 6 scale was 1.82, and only 15 percent gave it a 5 or 6 rating. Interestingly, students interested in the physical sciences were among the least enthusiastic. One question in the survey asked respondents to make a simple choice of the general academic field they liked best. The average rating given the SDI goal by the 121 students most interested in the physical sciences was only 1.5, matched by the 239 students interested in the humanities. Apparently this was not an issue that divided the scientists from the humanists. But the issue is powerfully shaped by other ideological commitments.

POLITICAL DISAGREEMENT ON MILITARY GOALS

Although it has frequently been pointed out that both Democratic and Republican administrations have taken aggressive international postures, politicians of the right generally give the military more enthusiastic support than do those of the left. How does political ideology influence support for Military goals, including SDI, and for other projected space missions?

A political question from the General Social Survey was incorporated in S1986A (Davis and Smith 1986):

We hear a lot of talk these days about liberals and conservatives. Below is a seven-point scale on which the political views that people might hold are arranged from extremely liberal (point "1") to extremely conservative (point "7"). Where would you place yourself on this scale? Please circle the ONE number that best indicates your general political views.

1 Extremely liberal
2 Liberal
3 Slightly liberal
4 Moderate, middle of the road
5 Slightly conservative
6 Conservative
7 Extremely conservative

Table 6.3 shows general support for the space program among three main groups, using data from the 1985 GSS. I have combined the categories of liberals and the categories of conservatives, to get political groups of comparable size. The space question is the same one considered in Chapter 1, asking whether current funding for the space program was too little, about right, or too much; again, I have combined "don't know" responses with the "about right" responses. Note that moderates are less likely than either liberals or conservatives to feel that funding was too little. The difference between liberals and conservatives is that the former are more likely to call for a reduction in funding and less likely to feel that current funding levels are about right.

Table 6.3. Political Orientation and Support for the Space Program, 1985

Those feeling funds for the space exploration program are . . .	Liberals	Moderates	Conservatives
Too little	13.3%	8.2%	12.3%
About right	43.2%	47.3%	51.6%
Too much	43.5%	44.5%	36.1%
Total	100%	100%	100%
Respondents	368	562	527

This table proves that political differences may be complex, and we should be prepared for differences in findings from one sample to another. Much of the American public lacks political mobilization, as low voter-turnout figures indicate. People at the ends of the political spectrum may include a higher proportion for whom political ideology really matters, and for many citizens a "moderate" response may be equivalent to a "don't know" or "don't care."

One could argue that moderates should be more supportive of the space program than citizens at either end of the spectrum. Liberals might want to shift the money spent on space to social programs. Conservatives may be against big government spending of any kind. Of course the topic of this chapter, military applications, may elevate conservative support and depress liberal support. But we should not forget that the domestically liberal Kennedy-Johnson administration was the one in which the space program flourished, and it was also the administration that got the nation involved in its first war of the space age.

Harvard students are undoubtedly among the most intellectually politicized in the nation, aside from members of organizations with explicit political programs. They have opinions about everything, and their intelligence multiplied by their youthful idealism produces a dazzling range of analyses of current social conditions. Every cultural innovation has reached the campus, and students come from every conceivable background, although disproportionately from the economic and intellectual elite.

On average, they are far more liberal than the average citizen, although all shades of opinion are represented. Consider the views of 947 respondents to S1986A who answered the GSS politics question and told us their sex. Fully 59 percent of these students said they were liberals, compared with only 25 percent of the GSS respondents in Table 6.3. Conservatives are only 29 percent of our respondents, but 36 percent of those in the GSS national sample. "Moderate" is the largest category for GSS respondents, but the smallest for Harvard students. Table 6.4 shows the percent supporting four aspects of the space program in each political category. I have separated the respondents by sex to show the substantial gender differences.

The bottom of Table 6.4 shows the numbers in each category. Because of the small number of women conservatives, partly reflecting the fact that most respondents were men but also an expression of greater feminine liberalism, I had to combine the three conservative categories for women. I also had to combine the "conservative" and "extremely conservative" men to achieve a large enough base for meaningful percentages. Across the board, the numbers of respondents are small, so minor, meaningless fluctuations in the figures are to be expected.

Table 6.4. Sex, Politics, and Attitudes toward Space Projects (percent)

7-Point Political Scale from the General Social Survey

	Liberal					*Conservative*	
	1	*2*	*3*	*4*	*5*	*5+6+7*	*6+7*
Research on defense satellite should continue.							
Women	12	19	40	64	—	78	—
Men	17	35	46	56	75	—	89
U.S. should build a manned space station.							
Women	21	33	40	56	—	49	—
Men	52	62	60	71	75	—	75
U.S. should set aside money for landing people on Mars.							
Women	30	39	46	52	—	47	—
Men	38	47	60	57	52	—	56
Space program should be manned as well as unmanned.							
Women	70	76	90	88	—	86	—
Men	73	85	90	89	93	—	98
Numbers of respondents in groups:							
Women	33	135	68	25	—	59	—
Men	52	141	132	84	130	—	88

The first space item was about SDI. The differences in opinion are huge. For example, only 12 percent of extremely liberal women wanted SDI research to continue. In contrast, 89 percent of the men at the conservative end of the political spectrum felt this way. Much of this difference is attributable to politics within each sex, but there is a hint that liberal men give slightly stronger support to SDI than liberal women. Greater male support is not obvious among moderates and conservatives.

The second space item concerns the Reagan administration's plan to build a space station. "Do you think the United States should build a permanently manned space station in orbit around the Earth over the next few years, or not?" In the 1988 presidential election, both candidates said they would continue this project, and Reagan was slow to adopt it himself, so it is not obvious that this is a Republican project rather than one with a broad consensus. However, my impression was that the university community saw it as an extension of Reagan's military buildup, even as an adjunct to SDI.

When Wernher von Braun (1952) proposed the space station project to the American public more than a third of a century ago, he stressed its potential military value. At the time, of course, he was preparing the Redstone missile for the U.S. Army in the midst of the Korean War, and he had long found that spaceflight advanced more rapidly when he could sell its real or imagined military advantages. But his promotion of the military value of the space station went to extremes of exaggeration. Early space pioneers, von Braun included, tended to minimize the value of small, unmanned vehicles. Despite his work on the V-2 and Redstone, he ignored the concept of long-range surface-to-surface missiles. And unmanned satellites were mentioned only as pathfinders to manned missions.

First, von Braun said, the space station would do military reconnaissance. Acknowledging the fact that the cameras themselves would have to be free-flying because the movements of men and machinery would jiggle the images, he none the less imagined them as satellites of the station, with daily visits by astronauts to retrieve film cartridges.

Second, the space station would participate directly in missile bombardment of hostile nations. The missiles would actually come from a small substation a couple of thousand miles behind the station in the same orbit. A missile fired from the substation would speed up as it dropped, coming under the main station that could track it and send radio commands to guide it precisely to its target. In part, von Braun expanded the role of human beings in space because automatic control devices were little developed when he wrote. But his lifetime purpose was launching humanity to the stars, and every military analysis or intermediate space project was merely a step toward that grand goal.

In Table 6.4 we see that the political left is much less favorable toward the space station than the middle or the right, and women show much lower levels of support than men. But the left shows more support for a space station than for SDI, and a majority of liberal men support the project. Note that the alternatives to saying "yes, the U.S. should build the station" included "no opinion" and failing to answer the question, as well as answering "no." Thus, our respondents are quite enthusiastic about the space station, even though their relative liberalism causes them to be suspicious of military implications of the project. In the next chapter, discussing the station as the first step toward colonization, we shall consider the reasons respondents give for this enthusiasm.

The two remaining space questions in the table were: "There has been much discussion about attempting to land people on the planet Mars. How would you feel about such an attempt —would you favor or oppose the United States setting aside money for such a project?" "Some people say the United States should concentrate on unmanned missions like the Voyager probe. Others say it is important to maintain a manned space program, as well. Which comes closer to your view." Modest political and sex differences persist on these two items, suggesting that militarization of space is not the only politically relevant factor.

Table 6.5 examines the connection between politics and eight space goals, the six of the Military class and the pride and prestige goals of the Nationalist order. In each case, liberals give lower ratings than conservatives, and moderates come in the middle.

Many observers of the American space program have been concerned that its civilian aspects may be swamped in the growing military aspects, with a consequent negative reaction from substantial sectors of the electorate. However, our respondents apparently do not see the space program as a mere extension of the Pentagon. One of the eleven government funding items in the GSS was "the military, armaments, and defense," and this item was included in S1986B. The average correlation between the 125 space goals and support for increased military funding is only 0.04, insignificantly different from zero.

Table 6.5. Defense, Nationalism, and Political Views.

	Mean Rating among Political		
	Conser-vatives	Moder-ates	Liber-als
109. The space program contributes to our defense.	3.59	2.24	1.61
37. There are great military applications of space.	2.93	1.74	1.04
96. A space-based anti-missile system, part of the Strategic Defense Initiative, could reduce the danger of war and nuclear annihilation.	3.28	1.71	1.16
21. The United States must develop space technology vigorously to keep up with the Soviet Union and other countries.	3.33	2.21	1.62
2. If America doesn't advance into space, some other country will.	2.67	1.78	1.25
30. Reconnaissance satellites help prevent war and nuclear attack.	3.63	2.79	2.35
67. The space program builds national pride.	3.00	2.67	2.59
26. The space program increases national prestige, producing worldwide respect for America.	2.74	2.04	1.47

One can wonder if the rejection of the Military goals spills over to the other space goals. The data show that it does not. The 117 goals not listed in Table 6.5 are essentially uncorrelated with political category. Thus, abundant evidence shows that respondents clearly distinguish Military goals in space from other goals.

THE SWORD AND THE SATELLITE

American culture sees a clear distinction between military and civilian uses of space, one drawn officially since the Eisenhower administration and reflected in the separateness of the Military class of goals measured by our surveys. However, the line between warlike and peaceful satellites in fact is unclear. Knowledge of the weather can greatly influence battle plans, as in the classic case of the Normandy Invasion, and civilian meteorology satellites become military ones merely with a change of personnel at the controls. Earth resource satellites provide the terrain contour information needed for accurate guidance of cruise missiles. The differences between communication and navigation satellites designed for military or civilian purposes are merely matters of detail.

Furthermore, the entire enterprise of spaceflight is built on a military foundation. Over half a century ago, leaders of the spaceflight social movement discovered they could not garner sufficient support from the general public, financial circles, or the scientific community. So they continued their journey to the planets via a military detour. With a certain amount of misrepresentation and wishful thinking, they sold their prototype spaceships as weapons, and the long-range missile was born. When international prestige became a valuable spoil of the cold war, the missiles became space launchers. The way to the Moon was a trek across a fierce but bloodless battlefield.

Now that the technology for near-Earth space missions and robot deep-space probes has been perfected, we can mentally separate military from civilian applications. But at root, the two are one. Critics of spaceflight, or of advanced technology in general, might well speak of the evils of the Faustian bargain. Goethe's Faust, it will be recalled, sold his soul to the Devil in return for knowledge and power. A modern German equivalent was Wernher von Braun, who unleashed the ICBM as a vehicle to the stars. Social theorist Oswald Spengler believed that each great civilization was based on a small, coherent set of ideas. Western civilization, he said, was *Faustian*. Thus, all of us have made Faust's bargain with the devil, not only von Braun. The essence of the bargain, stripped of the religious imagery, is the attempt to transcend the limitations of mortal existence through exercise of the intellect. The prime examples are science and technology. Except for very brief periods in a few ancient civilizations, science is the product of the West. Every society has technology, but systematic, vigorous, organized development of new technology is a Western innovation.

At the end, in Goethe's play, Faust in fact does not get condemned to Hell. Rather, his soul is redeemed for two reasons. First, Faust eventually devoted his intellect to making a better world through technological public works. Second, his indomitable spirit—"who strives forever with a will"—earns him salvation (Goethe 1965). Von Braun strove mightily for a great cause, and throughout his life he progressively developed an ever-deeper moral and philosophical understanding.

Eventually, perhaps, competition in space can replace competition on the battlefield, rendering spaceflight, in the famous words of William James (1911), "the moral equivalent of war." This phrase does not imply that the military is immoral. Quite the opposite. Despite his own pacifism, James acknowledged the positive benefits of warfare, including social solidarity, the motivation to excel, and rapid technical progress. In the modern age, he was convinced, the costs of war have become too great, and we must find an alternative source of these moral benefits. Especially if it is connected to general technological progress and economic growth, a space race can be a far safer and profitable challenge than an arms race.

Spengler (1926, vol. 2, p. 363) may have been foolish or melodramatic to claim, "War is the creator of all great things." But Wernher von Braun needed the war begun by Spengler's disciple, Hitler, to gain sufficient support to build the first prototype spaceship, the V-2. It is no tribute to the level of human evolution that the first voyage to the Moon was a maneuver in the cold war. Because we have not yet found the nobility within ourselves to explore the universe for the sake of wisdom, we must do so for sake of conquest. Modern social scientists find little merit in Spengler's analysis of history. Yet it is apt that Spengler (1926, vol. 1, p. 183) believed the perfect symbol of the Faustian, Western, ascendant soul was "pure and limitless space."

CHAPTER SEVEN

COLONIZATION OF OUTER SPACE

In the next century, after we have developed reliable means to loft large payloads into orbit, we will be ready to consider how far and fast our species should expand into the solar system. The spaceflight social movement has long proposed human habitation of space, and the technology necessary to transform this revolutionary dream into reality seems near at hand. Our questionnaire respondents had ideas and reactions to contribute to a discussion of interplanetary colonization, and this chapter will review them.

Twenty of the goals in S1986B concern aspects of colonization, and they clustered together in the block model analysis. The group is a distinct class, and the average correlation among the 20 members is 0.40, compared with an average of 0.24 between each member and the 105 other goals. This is a ratio of 1.67, and the ratio of squares is 2.78.

Ten of the twenty colonization goals have average correlations of 0.40 or greater with fellow members of the class. The highest average correlation, 0.50, is achieved by a direct statement of the central idea: "We could colonize the moon, Mars, and other satellites or planets of our solar system." In second place, at 0.49, is the proposition, "Space offers room for the expansion of the human species. Next, with an average correlation of 0.48, is "Space settlements could ease the growing problem of overpopulation." Tied for fourth and fifth place, at 0.47, are two general approaches to colonization, living in free space or on celestial bodies: "We could establish manned space stations, communities in space, and space cities." "We could find new worlds we can live on or transform a planet to make it habitable."

The second set of five goals, with average correlations ranging from 0.46 to 0.40, express the creed of colonization. "The Earth is too small for us, so we must expand off this planet." "Humans should spread life to other planets." "Our future ultimately lies in space." "We need an alternate home planet in case the Earth is destroyed by a natural catastrophe or nuclear war." "Eventually, interstellar travel could be possible, taking people to distant stars."

As we did in previous chapters, we can find orders of goals within the Colonization class by inspecting the highest correlation achieved by each goal. Four orders emerge, consisting of twelve, four, two, and two goals. The second is in some respects most representative of the group, but we shall begin, as before, with the order containing the largest number of goals. The first two orders are so powerfully correlated and their concepts are so similar that it might not be worth distinguishing them with unique names. But to keep them straight in our discussion, I will call them the Expansion order and the Colony-Building order.

THE EXPANSION ORDER

Table 7.1 shows the twelve goals of the first Colonization order, in three groups of five, four, and three items. The average correlation among them is 0.40, and goal 74, about expansion into the room afforded by space, holds the network of highest correlations together. The three parts of the Expansion order can be considered families of goals: Limitless Expansion, Departure from Earth, and Population-Food. They are somewhat distinct in their statistical associations, and the average intrafamily correlation is 0.46, compared with 0.38 across families.

The first goal in the Limitless Expansion family, "Space offers room for the expansion of the human species," is actually more subtle than it may seem. Respondents to the Spring 1986 survey meant not merely an expansion of the human population. They understood *expansion* in philosophical and moral terms,

Table 7.1. The Expansion Order (correlations)

	Average Correlation			45	92
	Intra-class	Extra-class	74 Expansion	Earth too Small	Over-population
Limitless Expansion Family:					
74. Space offers room for the expansion of the human species.	0.49	0.26	1.00	0.65	0.65
83. Our future ultimately lies in space.	0.41	0.28	0.57	0.52	0.44
94. Limitless opportunities could be found in space.	0.37	0.30	0.49	0.39	0.39
114. In space, we could create new cultures, lifestyles, and forms of society.	0.35	0.27	0.46	0.36	0.40
118. Mankind is bound to venture outside Earth and needs to know as much as possible in advance.	0.37	0.29	0.50	0.43	0.43
Departure from Earth Family:					
45. The Earth is too small for us, so we must expand off this planet.	0.46	0.21	0.65	1.00	0.64
111. We need an alternate home planet in case the Earth is destroyed by a natural catastrophe or nuclear war.	0.40	0.19	0.48	0.54	0.50
8. The space program could save the human race— by warding off a threat to the existence of our planet, for example.	0.35	0.21	0.39	0.39	0.41
46. Humans should spread life to other planets.	0.43	0.26	0.56	0.63	0.49
Population and Food Family:					
92. Space settlements could ease the growing problem of overpopulation.	0.48	0.23	0.65	0.64	1.00

Table 7.1. *(continued)*

	Average Correlation		74	45 Earth too	92 Over-
	Intra-class	Extra-class	Expansion	Small	population
7. Farms in space and advances in terrestrial agriculture aided by the space program could increase our food supply	0.35	0.21	0.43	0.33	0.49
70. The space program could help us control the weather, bringing rain to drought-stricken areas.	0.28	0.20	0.29	0.25	0.32

as well as demographic. There is a "need to constantly be expanding as a nation, as a people," and "expansion is necessary for freedom." "We need to expand our horizons," "expand our potential," "to someday get off the planet, to give mankind an expansive future. Earth cannot be home forever." "We should "expand reach for humanity" and "expand the usable portion of the universe."

"The race has got to keep growing long term. Thinking demands moving beyond terrestrial limits." This would mean "expansion of human species," "propagation of genes," and "concomitant weeding of the race." Here is a hint of perhaps the most controversial aspect of space colonization, the separation of the human species into Earth-bound and spacefaring subspecies, genetically different with the potential for mutual prejudice and bloody conflict. An "arrival of the fittest" ideology might emerge in space colonies, expressing contempt for humans left behind on the home world (Bester 1956–1957). But, future possibilities for racism aside, there is something grand in this vision. One respondent heard Browning's (1934, p. 51) poetry in the roar of the rockets: "Man's reach must exceed his grasp, or what's a Heaven for?"

The second goal in the family asserts, in the words of one of its utterances, "the ultimate future of the human race lies not here on Earth, but in space." Others used almost identical words, while for one respondent, the space program is "the *future*," and for a second, "*our* future." "The promise of space" "opens doors to the future." Colonization "is a vital step toward the future for the human race," and spaceflight "is necessary for our future to develop." We "will one day leave Earth and make our home in the universe."

The next goal was based on several other expressions of optimism. "The space program has potentials too widespread to attempt to enumerate." "There are incredible opportunities out in space." Others called them "endless opportunities," "limitless possibilities," "a myriad of possibilities," and "great potential." As one said, "Its the single most important endeavor mankind can ever embark upon."

Among the potentially radical results of space flight is the utter transformation of human society. Beyond the pull of Earth's gravity we could create "new cultures," thus achieving "increased cultural diversity." Establishing "societies on other planets" would lead "to hitherto unknown life-styles," "creating an alternate pattern of life that a lot of people will enjoy." This is to be expected from the space program partly because "knowledge and technology change life-styles and beliefs over time." Such changes may be essential "to develop alternatives to current ways of life that are careening towards nuclear destruction."

At least two features of extraterrestrial colonization would create diverse cultures. First, the material conditions of colonies would differ from each other and from Earth, requiring different modes of life. Second, the sheer fact of social isolation permits cultural drift and random innovation. Physicist Freeman Dyson (1979, p. 111) wrote, "It is in the long run essential to the growth of any new and high civilization that small groups of people can escape from their neighbors and from their governments, to go and live as they please in the wilderness. A truly isolated, small, and creative society will never again be possible on this planet."

The last goal of the Limitless Expansion family was based on only three utterances. One provided exactly the statement of the goal, and another seconded the conviction that "we will be in space." The third said, "Space is so vast, it would be crazy for us to be content just knowing about our own minute sector. We must at least know if there's anything out there to discover." Thus, this goal is not very well distinguished from pure optimism with a touch of the thirst to gain knowledge.

A family of four goals speak of Departure from Earth, beginning with the premise that the Earth is too small for us. But the utterances on which this goal was based really state a variety of ideas. "The Earth's too small," and "I'm bored with Earth," are not particularly compelling notions, but others say more. "The world isn't limited to Earth." "Our Earth is not inexhaustible nor unlimited." One suggested "we could leave this planet for good," and another wanted a "sense of not being trapped on a planet." One even asserted the space program should be continued "primarily so that humans can leave Earth."

Three respondents whose utterances contributed to this goal saw things from an evolutionary perspective. "The urge for man to expand is essential to his survival. We need space to expand into." "Getting off Earth is a big factor in continued human development." "Our race would eventually have to expand off this planet, assuming we don't blow ourselves up first. Space is the final frontier. If we don't cross that barrier, mankind is doomed to stay on Earth for the duration of its species life."

The theme of terrestrial destruction underlies the next pair of goals, establishing an alternate home and saving the human race. Respondents saw many reasons for needing an "alternate home," but the typical doom was "nuclear winter," "nuclear holocaust," "nuclear bombs," "nuclear war," and "nuclear death." "The human race dispersed over space is less likely to perish from a single disaster." Spaceflight must "get us off Earth as soon as possible. We are quite likely to commit racial suicide in the next few years." Spaceflight is "necessary for possible exodus from our planet," "security against Armageddon," "a guarantee for the eternal survival of humanity."

Other dooms would come "when we exhaust the Earth's resources," "poison the Earth so much that it no longer sustain life," or through "pollution," "ozone destruction," "infertility," and "major catastrophe." We have to study Mars as a possible home, because "If we don't look we can't leap if we eventually have to." But Mars would not help if "the sun supernovas." "The sun will eventually explode, and man will had to have found a new home by then." "When the sun begins to engulf the solar system, we can already have a place to go."

For some, all of us are guilty of despoiling the Earth, whereas for others the power elite is at fault. "We are very rapidly killing the planet on which we live; an alternate place to live is essential." It could provide "escape from Earth should those in power screw it up too badly." With varying degrees of seriousness, respondents considered genetic implications of terrestrial catastrophe. "With mankind on numerous planets, specieswide extinction will be impossible." "We would still have healthy germ cells available for the continuation of the species," and "somebody might survive the nuclear war without mutating into a cockroach." This is "the old theory of not keeping all your eggs in one basket."

The idea of defending this planet from doom, using the space program, was not well developed in the S1986A utterances. For example, no one mentioned preparations to destroy a stray asteroid or comet that might be found on a collision course with our planet. Two respondents spoke of "saving the human race," and another said the space program "might be our salvation." A fourth mentioned "thwarting something that might threaten the existence of our planet."

Only two respondents to S1986A contributed utterances to the final goal of the Departure from Earth family. One said explicitly, "Humans should spread life to other planets." The other merely mentioned "new places for life." Yet, despite its lack of popularity, this idea is an important one to have in our set, because it expresses a belief about the imperatives before the human species.

The Population-Food family of goals focuses on terrestrial problems, but because it postulates extraterrestrial solutions that would require colonization of space it deserves inclusion in the Colonization class. Thirty-one utterances from S1986A used the word *overpopulation*, and twenty-two others used the word *overcrowding* or some variant of it. Six spoke of "population pressure," and two cited the theory of Thomas Malthus (1888) that population tends to increase to the limit of available resources, stabilizing only at dreadful human cost. Other terms used were *population explosion, population problem, excess population*, and *lebensraum*.

Although one saw spaceflight as "solving the population problem," several were conscious of the limited impact that it could have, calling it "a partial solution" able only to "ease," "alleviate," or "relieve population pressure." One said that in space we "may eventually find room for our population, although I think we should be concerned with controlling it on Earth," and another joked about "colonization elsewhere when we get too promiscuous and run out of pills."

The idea of sending Earth's surplus population to other planets has been discussed seriously for more than half a century (Rynin 1931). To be sure, it is hard to imagine a program of colonization that could absorb the present growth rate of the human population (von Hoerner 1975). Whereas one respondent postulated "infinite room for human population growth," the potentially habitable environments of our solar system are far from infinite. Once colonization proceeds across interstellar space, the facts of geometry will impose limitations even if no other factors do. For example, an interstellar society that has a radius large in comparison to starship range can grow only at its outer surface, and all inner worlds are prevented from colonizing because all reachable environments are already inhabited (Bainbridge 1984). Of course, we are far from this point, and the fact that the economic investment per colonist is certain to be far above the per capita gross world wealth is daunting enough.

But, I think, the practical and logical limits often obscure the possibility that a combination of reduced fertility and modest colonization would be a far better solution than a complete stop to human growth. Peoples that collectively wanted to continue to grow would see hope of eventual great expansion through space. Furthermore, because some peoples of the world persist in having high fertility rates, zero worldwide population growth would effectively mean genocide or genetic suicide for the low-fertility groups, unless they had the opportunity of planting daughter societies beyond the confines of this planet.

Some students felt colonization provided "an outlet" or "space for future generations." "As the population of Earth continues to grow and natural resources shrink, space offers 'a new frontier'" and "expansion of human population beyond the bounds of an increasingly polluted Earth." Population problems are also "living standard problems." "It seems that our race is fast expanding; confining ourselves to this planet may prove fatal." In part, this is true because Earth-bound humans may fail to limit population growth, and the result would be incessant warfare. "If there ever is a lack of room for the population of the world to coexist peacefully, we will be able to expand our habitat regions infinitely should there be the possibility of existence on other planets." "Eventually, we're going to have to get off this planet. Resources low, population high." "We are already overcrowded on Earth. The most noble long-range goal is to help alleviate this by colonization."

The last pair of goals in Table 7.1 focus on farming in space and drought control on Earth. Logically enough, space farming correlates most strongly (0.49) with the colonization item about overpopulation. Everyone knows the world faces a problem of feeding a growing human population, and if we are to move people into space we must put agriculture there also. At 0.44 space farms correlate with "Some medical problems could be treated more effectively in the weightlessness of space." At 0.43 come two other colonization goals, establishing manned space stations and expansion of the human species.

Space farms link to the other goal in the pair at the 0.41 level, and this is the only coefficient as big as 0.40 the weather control goal achieves. It would be rather arbitrary to say one cannot go below 0.40 to look for connections to a weakly correlated item like weather control, and above 0.35 one finds several links to practical benefits of space, such as resources prospecting and solar power. Indeed, there are correlations with environmental protection items and those urging medical research and treatment in space, perhaps because all of these say something about human health. But raw materials from other planets also come in at 0.37, and the other items are not themselves so strongly tied to the practical benefits as to justify removing weather control from Colonization. Like space farms, it is a link from the worlds of outer space to this land we live on, suggesting that the activities of the colonists will benefit those left behind.

Several utterances mention farming in space: "growing food in space," "space farming," "crops produced in space stations," "the production of food supplies in zero gravity conditions," and—perhaps stated with tongue in cheek—"hydroponic intergalactic kelpfarms." Other respondents imagined ways space might aid terrestrial agriculture. "The sun's power could be controlled more effectively by satellites to increase food production on Earth." The space program might directly provide "benefits from research in the area of agriculture" or achieve "more efficient agriculture" through spin-offs. "Research tangential to the space program contributes to agriculture." Two respondents saw indirect benefits to the Earth, anticipating "new technologies for raising food in hostile environments" and "agricultural advances generated from learning to survive in harsh environments." The other goal in the pair was derived from just two utterances: "weather control" and "weather control for drought areas."

THE COLONY-BUILDING ORDER

The second order in the Colonization class, consisting of only four goals, is correlationally and conceptually the center of this chapter. Listed in Table 7.2, the goals say we could colonize space, but their clear implication is that we should indeed do so. Thus, they state a general program for development of interplanetary civilization and assert the value of colonization. First we would colonize the moon, establish free-flying cities in space itself, and begin emigration to the planets. Then perhaps with the power of interstellar travel, we could find new worlds and transform them to suit our evergrowing needs. To show the strong connection to the first order, Table 7.2 concludes with goals representing its three families.

Table 7.2. The Colony-Building Order (correlations)

	Average Correlation		64	24	6
	Intra-class	Extra-class	Colonize Planets	Space Stations	Distant Stars
64. We could colonize the moon, Mars, and other satellites or planets of our solar system.	0.50	0.28	1.00	0.68	0.58
24. We could establish manned space stations, communities in space, and space cities.	0.47	0.32	0.68	1.00	0.58
110. We could find new worlds we can live on or transform a planet to make it habitable.	0.47	0.25	0.66	0.55	0.54
6. Eventually, interstellar travel could be possible, taking people to distant stars.	0.40	0.28	0.58	0.58	1.00
Representative Expansion Goals:					
74. Space offers room for the expansion of the human species.	—	—	0.65	0.57	0.49
45. The Earth is too small for us, so we must expand off this planet.	—	—	0.62	0.56	0.49
92. Space settlements could ease the growing problem of overpopulation.	—	—	0.41	0.43	0.35

The average correlation linking these four colonization goals is 0.60, and their average correlation with fellow class members is 0.46. The average correlation linking these four to the twelve Expansion goals is also 0.46, substantially higher than the 0.40 connecting members of the larger order.

The first goal in the Colony-Building order was based on a wealth of material from S1986A. Many respondents foresaw "having human beings living and working in space over long periods of time, even their lifetimes"—"humans living on celestial bodies other than the Earth for long periods of time: colonization." "In the long run, there may be human populations born and raised on other heavenly bodies." "We should colonize the rest of the system and eventually (hopefully) the stars," to "extend our horizons to outer space with the hope of one day living in space." "Space exploration opens with new possibilities for our own lives—living in space," "expansion to new frontiers (new human habitats, new resources)," and "alternative places for people to live and work."

The most common term given for this was *colony,* and ninety-six utterances used some form of this word. Eighteen utterances used *settlement* or a variant of it. There is an interesting difference of connotation between these two that I noticed at an international space conference. Citizens of Third World nations and their sympathizers tend to consider *colony* a bad word, because in their experience it refers to the oppressive colonial empires that England, France and other European powers had created. But for citizens of the United States, it has positive connotations, reminding them of the heroic, original thirteen colonies from which their nation grew. One respondent urged, "Think of the 16th century Spanish and Portuguese explorers," suggesting that the colonization of space was a natural consequence of exploration like theirs. Other words associated with the idea in S1986A included "expansion," "emigration," "habitation," and "populate."

The concept of *lebensraum* emerged in statements like, "plenty of elbow room," "living space," "providing extra space for humans," "expanding populatable areas," and "this world needs new land." There could be "space colonies for living and perhaps growing food," even "self-supporting" ones. A student who professed to be "just kidding" suggested "permanent space-dwellings" could provide a "homeland for Palestinians."

One wrote of "colonization and resource potentials," saying, "Both would enable us to preserve the integrity, peace, and beauty of the Earth." Another combined colonization with spin-offs. "I think that the settlement of other worlds would be most significant—not simply for the additional space, but because the technology needed for such settlements could be applied to improve the quality of life for everybody (especially, Third World peoples) on Earth." Two students linked colonization with present mismanagement of our planet. "Get off Earth when we exhaust resources." "The Earth will eventually run out of resources, especially given this consumer–throw-away culture we see as the height of civilization."

Several sites were suggested for colonization: "outer space," "the Moon," "Mars," "other planets," "other moons," "celestial bodies," "the solar system," "other solar systems," "the universe," "other places," and "elsewhere." One said, "in the very long run (1000+ years), other stars," and another even suggested the presently implausible "other galaxies."

Although most of the attention was directed toward other planets, several respondents imagined the establishment of manned space stations and communities floating in space. Design studies have firmly established the technical feasibility of such space settlements (Johnson and Holbrow 1977), and the question becomes one of economics or social value. Respondents to S1986A offered positive evaluations: "The possibility of space stations is valuable as not just a practical but an exciting prospect for future living." One apologized for sounding "like Arthur Clarke" in citing "permanent space stations" as a goal. Others saw "expansion of population onto satellites," "perhaps a space city" or "communities in space." When asked what might be a significant long range result of a vigorous space program, one simply answered "L-5." This refers to the proposal of physicist Gerard K. O'Neill (1977) and the L-5 Society to establish a city in the orbit of the Moon.

Closely related to the simple idea of planetary colonization—indeed perhaps just a variant of it as indicated by the very high (0.66) correlation between them—is the goal of finding new worlds or transforming a planet to make it habitable. The distinction here is the goal's implication that a perfectly suitable site for a colony is not currently known, and thus one must be either located or created. Respondents spoke of finding "hospitable planets," "livable planets" "habitable planets," and "new worlds to inhabit." One spoke of "Eden," and one thought the space program might find "another environment better than our own that people can inhabit peacefully."

Other respondents spoke instead of "transforming non-Earth environments into habitable lands," what is technically called *terraforming* (Oberg 1981). One person felt "a way to inhabit other planets" "would be very significant." Others wrote of "making Mars habitable to humans," and "preparing other planets for colonization."

To find new worlds we must go beyond this solar system. S1986A respondents saw "the possibility of interstellar travel and discovery," perhaps even "movement between galaxies." For one, "interstellar travel would be amazing," whereas another wanted "mankind to expand from the solar system and achieve immortality." There were no lengthy comments on the feasibility, but one respondent distinguished two variants: "family space travel (travel over generations) *or* speed of light method of travel."

THE RESOURCES AND ANTIPOLLUTION ORDERS

Two small orders, each consisting of a pair of goals listed in Table 7.3, conclude the Colonization class and may be called *Resources and Antipollution*. One consists of two statements about raw materials and mineral resources, and the other is about disposal of wastes and removal of polluting industries into space. The average correlation linking goals in one with goals in the other is 0.39, and the resources order is most strongly connected to Colony-Building with an average of 0.42.

The goals of the Resources order, raw materials and minerals, are nearly identical in meaning, as proven by the overpowering 0.66 correlation between them. Both are connected to a goal in the ecology satellite of the Technical cluster: "New fuels found in space or the development of fusion power in space could help solve the Earth's energy problem." New mineral resources has a correlation of 0.55 with new fuels, and raw materials achieves 0.48. These two goals could be placed in the ecology satellite, but as Table 7.3 shows, they also have solid connections to colonization.

They also anticipated the "discovery of raw materials useful on Earth," whether "from the Moon" or "from other planets."

"There are incredible resources out in space," and these "abundant resources" could substitute for "depleted resources on Earth." "Space exploration is vital to human survival as our resources are slowly diminishing." "In the long run, the resources of outer space will be very important to human life. In order to explore there, we must be developing space travel now."

Table 7.3. The Resources and Antipollution Orders (correlations)

| | _Average Correlation_ | | 121 | 73 | 59 |
	Intra-class	Extra-class	Raw Materials	Safe Disposal	Industrial Pollution
Resources Order:					
121. We could use raw materials from the moon and planets when natural resources are depleted on Earth.	0.39	0.21	1.00	0.43	0.41
40. We could find new mineral resources on the Moon, Mars, or the asteroids.	0.38	0.23	0.66	0.36	0.34
Antipollution Order:					
73. The Moon or the sun could be used for safe disposal of toxic materials and nuclear wastes.	0.31	0.17	0.43	1.00	0.59
59. We could preserve Earth's environment by moving the most polluting industries into space.	0.29	0.15	0.41	0.59	1.00

S1986A respondents foresaw "mining in space," "mining Mars," "mining meteors," and "mining of the moon or asteroids," which might provide a "solution to mineral shortages," I was impressed that the asteroids were the most frequently named site for such mining, mentioned by ten respondents. One observed, "We might learn to mine asteroids void of life, rather than destroying the life and beauty of our world."

The last two goals in Table 7.3, the Antipollution order, seek to protect the Earth by transporting dangerous materials and industrial processes beyond the biosphere. The idea of using the moon for disposal of nuclear waste was popularized in the otherwise technically absurd television program, *Space 1999.*

Utterances in S1986A clearly stated the idea of using space for "dumping and disposing of hazardous waste," "toxic wastes," "garbage disposal," and "nuclear waste disposal." Although one respondent suggested "storing them on the Moon," another proposed "sending nuclear waste to the sun." Equally clear statements identified the twin goal of "removal of industrial processes from biosphere," getting the "more harmful industries off of the Earth" "to preserve Earth's environment." As one put it, "Let's put industry and ecologically damaging chemicals in space and keep the Earth beautiful."

Like the two previous space goals, the concluding pair in Table 7.3 seem connected to the ecology satellite of the Technical class, described in Chapter 4. To examine this connection more closely, Table 7.4 shows the correlations linking these Colonization goals with the four ecology goals. The coefficients vary over a wide range, and it is clear that the eight do not form a cohesive cluster of ecology-resource goals. But, the ties are substantial, and it is worth recognizing the concepts several of these space goals have in common.

Three of the four ecology goals enjoy substantial correlations with the four items in the Resources and Antipollution families of the Colonization class. The exception is the one about nuclear winter, which may be more an expression of political attitude than of real enthusiasm for expanded human presence in space. To use space to improve our world's ecology and energy use, we must develop a bridgehead above the atmosphere, the first step in serious colonization.

Today's Earth satellites do not require an extensive human presence in space, and at present no nation has the capability of routinely visiting synchronous orbit to service the communications and meteorology satellites stationed there. Only if we expand exploitation of near-Earth space do permanently manned stations and highly developed orbital transfer vehicles make sense. But as orbital operations expand, an extensive infrastructure must be built, the first steps toward colonization.

Table 7.4. The Bridge from Colonization to Ecology (correlations)

	121 Raw Materials	40 Minerals	73 Waste Disposal	59 Polluting Industries
63. Solar power stations in orbit could provide clean, limitless energy to the Earth.	0.34	0.38	0.29	0.24
72. New fuels found in space or the development of fusion power in space could help solve the Earth's energy problem.	0.48	0.55	0.30	0.26
62. From space, we could find new ways to control pollution and clean up our environment.	0.31	0.34	0.28	0.32
112. Through space research we learn the true extent of devastation that nuclear war would bring: nuclear winter.	0.12	0.15	0.06	0.09

In particular, solar power satellites have often been mentioned as the justification for establishment of space cities and factories. Substantial crews and equipment may be required to build them, and some writers have suggested using materials from the Moon for structural elements or even more sophisticated components.

When respondents express enthusiasm for asteroid mining, solar power satellites, or protecting the Earth's environment, they link colonization to the betterment of conditions on our home planet. These goals draw benefits down from colonies. The other side of the equation must be the reaching upward from Earth toward colonization. In questionnaire S1986A, a question about the current space station project and four items about Mars garnered student's views on the possible first steps in colonization.

IMAGES OF THE SPACE STATION

At a time when there had been much public debate concerning NASA's plan to establish a space station (O'Leary

1983; cf. Compton and Benson 1983), it seemed important to include in S1986A the item, "Do you think the United States should build a permanently manned space station in orbit around the Earth over the next few years, or not?" In the previous chapter, we saw that political liberals were less likely to favor this project than political moderates and conservatives. Altogether, 124 (38 percent) of 324 women and 435 (67 percent) of 654 men did so.

Those who wrote comments most often mentioned the scientific benefits of a space station. "A manned space station is vital for conducting extensive long-term experiments in space. It is the only place where there is negligible gravity for significant amounts of time, and the only place out of the Earth's atmosphere." "The microgravity experiments possible" could determine the "effect of gravity on developmental biology and the behavior of substances," "useful for drug research," and let us "learn about human behavior in zero-G for long times." The space station "would provide a place for an impressive telescope" which "would permit the study of the outer atmosphere as well as coronal activity without any obstacles," and it would be "critical for astrophysical research." "It would provide a platform for increased and better Earth research," including "better weather studies."

Also frequently mentioned were applications for industry. The space station "will facilitate the advance of technology," in such areas as "growing crystals," "microchip manufacturing," and "drug manufacture." "Some illnesses can be cured in weightlessness," and there are many "possible advancements that could be made in medicine." We "need it for potential solar power," and "this links ultimately to the development of new fuel sources." In the short run, "commercialization of space" needs "government subsidy to ease entry." "We can start new industries that way" and "increase GNP." "In order to get the best benefit, we need to use our technology to make money, as well as explore," and the space station would be the "first step toward manufacturing in space."

Respondents were aware that the space station "would make ongoing research much easier" and facilitate "orbiting of satellites" and "maintenance of satellites." "Instead of spending millions on temporary stations, why not a more permanent one?" "Then we would not have to worry about dangerous takeoffs and landings" or "to send up so many shuttle missions, which are costly, whereas a station requires only an initial investment." "It could reduce engineering costs in the long run" and make "experimentation and space travel cheaper." "It would lessen greatly the cost of sending rockets to other planets."

The space station was seen as having implications for international competition. "It is worth notice that the Soviet Union presently leads the world in developing permanent space stations, having had Salyuts in orbit for roughly a decade." "Russia is so far advanced in that area," "so we need to keep up." "The U.S. should establish a presence in orbit." "The Russians have one, so we need to keep up." But seven respondents argued that it should be done "in conjunction with other nations" as "a cooperative venture." "If there is going to be a space station, then it should be a joint program between the U.S. and U.S.S.R., and other countries. For only the U.S. to have one would be a threat, and space exploration should be a joint human venture, not a jealous race."

Those who saw a military potential for the space station were divided over whether this was good or bad. "It would have useful military functions," "for military research" "in preparation for the Star Wars program." "Can put missiles in them. Better to have first strike capability." In contrast was the view the space station should be built only "if we could guarantee its uses to be limited to medical and other peaceful purposes, not "used as a weapons station, "not to find targets in Russia." "If nonmilitary," the station would be "of much more value than SDI." "I'd rather see the tax revenues go for space stations than for missiles or conventional weapons."

For many respondents, the space station "is the next logical step in space exploration." "This is a necessary part of the shuttle program. The two will someday be a natural combination." "It provides a good staging area" "as a base for further space exploration." "This is a giant first step towards being able to really use resources away from Earth." "Space is the next frontier for human development. The closer we get to establishing a permanent foothold in space, the further we move from potential stagnation." "A necessary prerequisite for useful manned space exploration," the space station is "the first step towards the stars."

Further, the space station would be the "first step toward habitation of space." It would offer "experience in space construction" and "could over time (with parts being ferried up by shuttle) be the launching point of the L-5 colony." Thus, an Earth-orbiting space station "would be a constructive step towards colonization and exploitation of the resources of our moon and other solar system planets and moons."

In sum, the space station was seen as a reasonable "goal to shoot for," an important part of "the future of the planet," perhaps "the doorway to the future." "If this objective can confidently be achieved without huge risks, then why not? Without the courage to 'take the next step' we'll never learn anything." "Despite the Challenger accident, NASA has a pretty good safety record. With proper precautions, then such a station could be both profitable and safe." "A permanent space station would encourage research and space development as well as save money in the long run. A station will be built—better by this generation than the next." "I would consider a permanent station an important step toward the opening of space for the 'common man.'" "It would create a sense that space, at least so far as an orbiting station, is not so far away nor so alien as it might seem."

EXPLORATION AND COLONIZATION OF MARS

A manned mission to Mars has been a perennial dream for space enthusiasts, and the feasibility of such a voyage was demonstrated long ago (von Braun 1953; Ley and von Braun 1956). Other major space goals may be targeted first, certainly the space station but most probably a scientific base on the Moon (Mendell 1985). Personally, I find Rhea and Iapetus very attractive for manned exploration, but these moons of Saturn are so far away that they can be examined only by robot probes for the near future (Morrison 1982). And of all the other worlds of the solar system, Mars is most hospitable (Murray, Malin, and Greeley 1981).

As Chapter 1 noted, in 1969, the Gallup Poll included the following item in a national survey: "There has been much discussion about attempting to land a man on the planet Mars. How would you feel about such an attempt—would you favor or oppose the United States setting aside money for such a project?" I included this question in S1986A, and noted previously, respondents at the liberal end of the political spectrum were slightly less enthusiastic than others. Altogether, 137 (42 percent) of the 324 women and 344 (53 percent) of the men were in favor of an expedition to Mars.

In the open-ended section of S1986A, I included two related items. "What is the most important reason for exploring and settling the planet Mars?" "Can you mention another reason for exploring and settling the planet Mars?" By combining responses to these two items with replies to the first one given by supporters of a Mars expedition, we can sketch the image American culture has of the value of the project.

The most commonly supposed benefit to come from exploration and settlement of Mars was knowledge, including scientific discoveries and practical information, but it is hard to count precisely the utterances in this category because it is so diffuse and multifaceted. Topics mentioned included "astronomy," "biology," "cosmology," "geology," "planetary evolution," "study of atmospheric change," but the most common words used were simply "knowledge" and "science." "It would be a fascinating scientific experience—a whole new world." An expedition would permit "examination of geological aspects of a different environment, under different ecological and atmospheric conditions" from those of Earth.

"The configuration and development of Mars is largely a puzzle." "Mars holds clews about the origin of the solar system" and "is important in our studies of the physics of the solar system." It is an ideal site for "study of processes of geologic change," allowing us "to develop an accurate physical history for another planet—apply recent developments in geophysics and plate tectonics." We would gain "knowledge of solar and terrestrial evolution," learning, for example, "why the planet dried up."

Although this knowledge "would not be geocentric," it would help us understand "the formation of the Earth." "By understanding Mars we can understand the Earth better." Compared to the Moon, Mars "holds more secrets about our own planet." "Comparative insight" will come from "examining and learning from a planet that is both similar and different from the Earth. It may help man to better understand his own world and teach him how to take better care." This would include "what makes life on Earth possible and life there impossible" and give "a greater appreciation of life on Earth and the origins of life in general." We could "learn about geological processes on Earth," "how the Earth was formed," "and why the Earth and Mars evolved differently, e.g. plate tectonics, different atmospheres."

Some students were aware that the Viking landers had failed to find evidence of life (Cooper 1980; Ezell and Ezell 1984). "A manned mission could settle the question of whether or not Mars has life, a question left open by Viking." "Perhaps we'll discover life after all, since the landers could not explore." "Maybe there's another form of life unknown to us." Some students joked about "enslavement of Martians" or wanted to "find out if Martians really look like the Martian on Bugs Bunny" and to "check out the Martian women." Others hoped "we can dispense with the 'little green men' myth" and thought life on Mars might be in the form of "viruses." "Mars may once have held life, which may have become extinct, and we probably would like to know about that," "discovering how life on Mars developed and died."

The second largest category, 173 utterances, referred to the "untapped natural resources" of Mars, 50 of them using the word *minerals*. Indeed, given economical means to exploit them, the red planet possesses varied geological environments that can be assumed to have abundant resources (Carr 1981). One to three respondents mentioned each of the following: "energy," "fuels," "gases," "iron," "metal," "power of sun," "strategic metals," and "water." Most respondents said nothing about where the resources would be used, whether on Earth or Mars, but mining implies mining settlements and support systems that only substantial colonization could achieve. A few specifically wrote about "resources on Mars which are transferable to Earth," "resources for our overcrowded blue marble."

There were also tentative hints that Mars might offer "economic benefits," other than raw materials, to "business," "industry." "Lower gravity may be useful for high-tech manufacturing." Mars may have other "commercial" uses "to create wealth," "to create jobs," and "to expand the human economy." Twenty-one said something about farming on Mars, which could become "a place to grow food for the Earth's billions." However impractical the idea, "Mars could be set up as a resort planet for weary space dwellers to vacation to," according to six Harvard students. Poorly-informed romantics praised "the view of the canals" and "the view of Earth on a summer's eve." More reasonably, the Mars project "would probably create many spin-offs." "Getting there would require us to invent new technology." Living there would "force us to develop new technologies to cope with Mars," which would generate a "windfall into society."

Seven respondents favored the Mars project as an alternative to military spending. "This would be far more valuable than the development of defense satellites," "more realistic than SDI," and offering "greater potential for learning about universe as a whole." "How about using some of the money already wasted on the thousands of unneeded nuclear weapons?"

Many students would "like to see a joint U.S.-Soviet project land on Mars," and three of them cited Carl Sagan as their source for the idea (cf. Joëls 1985; Collins 1988). "If the program were carried out with the Soviets, it might improve relations." "It would be very good for world peace should the U.S.A. and U.S.S.R. attempt a joint mission," and "a good way to vent imperialistic tensions." "We must forget insignificant national conflicts when we move beyond the Earth." To reduce "costs and give a sense of cooperation," "it should be done with participation of other nations." A United Earth expedition "would foster international understanding," "divert attention from our differences and emphasize our similarities," through the "unification of a large, common goal."

In contrast, others wanted to keep the expedition exclusively American. "It would be an achievement of such magnitude that it ought to be done by Americans." "*DO NOT* cooperate with the U.S.S.R. on this. This is the first step to further our settlement of space." The "Russians are going for it," and "we must relieve the siege it's under by the Reds!" A U.S. expedition would "keep America first in progress and technology," "beat the Soviets," and "maintain world standing." "Landing Americans on Mars would be a great symbolic achievement." "Such a project would build up national pride, as the space program of the '60s did when the unified goal of the nation was a landing on the Moon." Indeed, a new space race would "revitalize scientific competition among nations" to the benefit of all. "The U.S. likes a big show—man on Moon, reusable shuttle—why not man on Mars?"

An expedition to Mars and Martian settlement drew a substantial number of emotional justifications. Twenty-three respondents mentioned "curiosity," and others cited "fun," "excitement," "kicks," "pleasure," "the experience," "wanderlust." Exclamation marks abounded. "This is the exciting part!" "That's an incredibly exciting idea. Go for it!" "Adventure!" One took it personally, "sounds fun, where do I sign up?" Another wrote it would be "exciting for the country." "Remember the feeling of landing on the Moon? Let's recapture that feeling!"

As we might expect from the hundreds of utterances that produced the space goals discussed in Chapter 5, students also gave idealistic reasons of many kinds for going to Mars, not just nationalism and internationalism. "It's the next step in our evolution as explorers." We must go for "sake of the advancement of mankind," to "keep progress going," and "progress and daring new steps are vital." "Man by nature wants to expand, first to other parts of the Earth, then to other planets." "Man must always reach for the stars, otherwise humanity will become stagnant."

Twenty-one said simply, "Because it is there." Others saw it as the traditional target of spaceflight "It's always been one of mankind's dreams." Going to Mars would be "immediately symbolic" "fulfilling a dream," satisfying "the Hunger of Imagination" and "the spirit of exploration." We could gain "new perspectives on existence" that "would rejuvenate literature" and "invigorate the Earth with accounts that explore the range of human sensory perception." "Our idea of how everything works is confined to the setting of Earth. Exploring Mars could open up new ideas and ways of viewing things." We should go now, "because we can!" "One day men will be on Mars, just as they have been on the Moon—why not now rather than later?" "We should have the same zest and interest toward Mars as we did about the Moon." "To let this generation go unfulfilled would be a tragedy."

For some, "the conquest is the key. Information gotten is not so huge as to justify the cost." We should go "for the psychological benefit of 'conquering' another major planet of the Solar System." "It's a new frontier to be conquered," for "the human ego," "the pioneer spirit" and our "manifest destiny." Settling Mars would "give people with adventurous spirits an outlet—a *frontier*." One praised "the romantic aspect of it—settling the final frontier—cowboy stuff." "Man lives in such a small part of the universe—going to the Moon, 250,000 miles away, is stone skipping. Let's go to Mars."

In many ways, respondents explained that "pride is an important social factor." "This is how we will be judged by future civilizations." "It's ridiculous to go back to our pre-space age." "Man's conquest of space is a *noble* endeavor." We should explore Mars "because it's a challenge" and "to show we can do it." "It would be more than justified by the achievement of actually landing human beings on another *planet*. Can you imagine?" "Get a man on another planet, and the future is here." "To reach another planet would be a real feat for mankind." "What the hell—let's boldly go where no man has gone before!"

As they did in response to general questions about the space program, respondents often suggested that "new territory for man" could "ease crowding on Earth," providing a partial solution to "overpopulation." This is, as I have suggested, a very complex issue, although most respondents seemed merely to think Mars adds to the space available for a growing population. Some saw it as providing "a safety valve for overpopulation" or a "vent for exploding population," suggesting through this metaphor that Mars could absorb all demographic increase, leaving Earth with a stable number of inhabitants. One wrote of "easing the population burden and easing corresponding racial and political tensions caused by overcrowding." Another imagined that mass emigration to Mars would "leave room for the expanding human population. Then the entire Earth can be cleaned up and used for food production." But optimism was not universal. One agreed in the general principle that Mars might relieve terrestrial overcrowding, but "this may not be feasible for several hundred years." Another was ready to give up hope for Earth and transfer it to Mars. "If this planet becomes overcrowded, at least a new civilization could start there."

I was shocked to see again how many respondents felt destruction of the Earth was a real possibility. Recall that one of their general goals for the space program said, "We need an alternate home planet in case the Earth is destroyed by a natural catastrophe or nuclear war." Fifty-six respondents proposed Mars could be "an escape hatch" "to preserve humanity after nuclear war." Either we would "get humans off Earth before we blow it up," or Mars would be "a backup to retreat to" "so that when we nuke our own planet to pieces, we have somewhere to go." Although we could move to this "alternative living environment should Earth become uninhabitable," "if we survive," it might be safer to establish a colony before and thus "divide population so we aren't destroyed in a nuclear war." This would "double our survival chances," and Mars "could be a refuge for those trying to escape an all-out nuclear war," "as in *The Martian Chronicles* by Ray Bradbury."

Three respondents specifically cited natural disasters. "When the ice ages come and many if not all humans are displaced, Mars, if it could be adjusted for humans, would be a shelter." We might need to emigrate "in case the mass extinctions are a result of asteroids hitting the Earth every 26 million years." One even considered the sun's expansion in the far distant future. "Mars is further from the sun and will probably last longer than Earth."

Happily, many students have a more positive view of colonization. "It might be economically feasible to live on Mars," which has "the most favorable conditions for human existence of any other planet." "Trying to put a possible space station there," "could prepare the way for large-scale settlement." "I think colonization is a key to the future of the human race."

On Mars, we could "set up a utopia free of this world's problems." Mars could be declared a "nonpolitical zone" and made a "nuclear bomb-free environment, a world that could live without having to fear enemy missiles," "a human colony safe from Terran intrigues." It would "give humanity a more homogeneous and assimilated place to live. Mars would be a 'happy' country, rather than divided and warring." "The Earth would have to work as a single unit in its relations with this new settlement, which would be conducive to peace." The first Martians would begin by "framing a new society," a "better lifestyle," an "alternative to Earth living." They would "create a new land of opportunity for adventurous scientists, businessmen," and a "new start—for the lucky."

"Mars is a good experiment for testing the possibility and feasibility of people living on other planets." We should try living on Mars to see "if we can adapt to that environment" and "to see if life is perhaps changed by inhabitation in another location." In colonizing Mars, "we could learn vast amounts about living in space," "experiment with extraterrestrial settlement," "develop the skills and technology to explore and settle other locations." On Mars, we can "learn how to go farther."

Six respondents hoped ultimately we would be capable of "transforming Mars into a liveable planet" (Oberg 1981; Allaby and Lovelock 1984). "It has an atmosphere—it can be terraformed." The ability "to build up the Martian atmosphere so humans can grow crops and live," would give us "progress in terraforming which we could apply here." "If we could reengineer Mars into a more Earthlike world (terraforming), we could easily develop means to repair our damage to our own world."

Near-term benefits to the space program itself were frequently cited. A Mars expedition would be "an excellent way to develop heavy-launch capability," "to advance the technology of manned spacecraft to farther and farther distance," and to "improve space technology in general." "It would bring more confidence in the space program," more "public support," and more "monetary support." It would end "the national retreat from manned space travel that started in 1972." If the Russians responded in kind to an American space challenge, the expedition would achieve "revitalization of the 'space race.'" Why set a specific target like Mars, rather than merely increasing general space development? "Mankind needs a goal to sustain any effort."

"Mars seems the next logical step" "in space exploration." "A Mars flight is the next step beyond the Moon" and "a prelude to further exploration." "We'll learn a lot by this short excursion before we move on to the longer ones." "We have to start somewhere" and "begin the process of moving off of our planet." The Mars mission will "get us a start moving out into and exploring the rest of the universe." "It is a good takeoff point for exploiting the asteroid belt," and a "stepping stone to Jupiter" and "the outer planets."

Mars will become "a base for further exploration by manned vehicles," "a first step in our branching out into the universe."

But, finally, it is impossible to list all that we might gain from Mars, because "unforeseen benefits often result from such exploration." "Who knows what Mars has until we go and look?" "Who knows what unexpected advantages could come out of it?" "Who knows what we might find there? High risk, possible high return." "We may and hopefully will find discoveries unexpected previously." "Really, it's just an opportunity for new experiences and discoveries. Who knows what will result? When you travel abroad you don't know the definite results, because it's a new experience which adds to you as a person."

A RACE OF COLONISTS

Not just Americans, but all humans are descended from colonists. From our ancestral home in Africa, we have spread out to conquer one planet, and now others beckon. In response to the question about whether we should prepare an expedition to Mars, one replied, "I would like to go there myself." A second said, "I am studying to be an astrophysicist and hopefully an astronaut. I want to be aboard the first manned mission to Mars. No kidding." An item in S1986A placed the issue in a more personal context for all respondents, asking, "If you were asked to go along on the first rocket trip to the planet Mars, would you want to go or not?" Of 324 female respondents, 154 (48 percent) wanted to go on the first mission, while 425 (65 percent) of the men expressed personal interest in this risky voyage.

When it comes time to plant a settlement on the planet Mars, apparently there will be no difficulty finding qualified volunteers. To be sure, the students could not have taken the Mars trip question very seriously, because they know that the first expedition is many years away. But their responses reflect considerable thought and many of them would indeed have much to contribute to the colonization of a new world.

If the Technical class of goals concerns immediate practical benefits, and the Idealism class concerns timeless or abstract values, Colonization looks into the future with a mixture of rationality and imagination. If we invest in further development of space technology, we will be able to establish viable colonies on the Moon, Mars, and possibly outer moons of Saturn. Motivations for very early stages, such as construction of solar power satellites, might come from ordinary terrestrial needs. But to take the great leap to the planets, we must find transcendental ideals that will drive us outward. This is a prime question for the human future. Are we to remain forever on this tiny globe, lost in an immense universe, or can we fulfill the revolutionary values of the Colonization class?

Ray Bradbury has asked why we should forswear our terrestrial obsessions and voyage outward, and he provided an eloquent answer. "Because, wouldn't it be terrible to wake one morning and discover, without remedy, that we were a failed experiment in our meadow-section of the Universe? Wouldn't it be awful to know that we had been given a chance, a testing, by the Cosmos, and had not delivered—had, by a loss of will and a flimsy excuse at desire, not won the day, and would soon fade into the dust—wouldn't that be a killing truth to lie abed with nights?" (Bradbury 1977, p. 10). Instead, how fulfilling to hope that a thousand year hence the citizens of a dozen worlds will honor us as their ancestors and the founders of their societies. This hope demands fulfillment.

CHAPTER EIGHT

COMMUNICATION WITH EXTRATERRESTRIAL INTELLIGENCE

A powerful wave of technical and popular publications has recently asserted the feasibility and desirability of a search for extraterrestrial civilizations. Although it is recognized that we need to understand the social sources of support for this great project (Molton 1978), discussions of them have tended to be purely conceptual (MacGowan and Ordway 1966, p. 179–272; Sagan 1973, p. 85–187; Stephenson 1977; Singer 1982), and the few empirical studies have been historical essays (Bell 1980b; Dick 1982). Because spaceflight was achieved by a social movement, rather than as the inevitable consequence of technical progress, it is quite possible that the decisive factors in achieving communication with extraterrestrial intelligence (CETI) will be social as well.

Two of the 125 space goals, items 28 and 51, concerned extraterrestrial life: "We could learn much from contact and communication with intelligent, extraterrestrial beings." "We might find alien forms of life on other planets." Each of these correlated more highly ($r = 0.64$) with the other than with any of the remaining 123. But seven of the other goals achieved a correlation of 0.40 or greater with one of this pair: "Eventually, interstellar travel could be possible, taking people to distant stars." "We could establish manned space stations, communities in space, and space cities." "We could find new worlds we can live on or transform a planet to make it habitable." "Humans should spread life to other planets." "We could colonize the moon, Mars, and other satellites or planets of our solar system." "We should explore the unknown." "Space settlements could ease the growing problem of overpopulation."

Thus, contact with ETs appears connected with colonization and other visions of distant possibilities, yet it is extremely popular as a potential aim of the space program. We shall not examine the relatively limited utterances from S1986A on which the two ET goals were based, because we have better data that will form a major part of this chapter. And we shall begin entirely outside the framework of the 1986 survey project, in earlier polls about extraterrestrials.

MARTIANS AND OTHER POPULAR MYTHS

The earliest poll data we have on extraterrestrials were collected in the aftermath of the great Martian invasion scare of 1938. As a special Halloween treat, Orson Welles and his Mercury Theatre broadcast a radio dramatization of *War of the Worlds* by H. G. Wells. Although introduced as fiction, the drama simulated news bulletins of a real Martian invasion, and a substantial number of listeners apparently believed they were in imminent danger. The extent to which an actual panic occurred remains in doubt, and the episode might as well be described as a mass media craze in which journalists reported evidence of panic to have an exciting story to tell (Bainbridge 1987a). In any case, the Columbia Broadcasting System did a poll, and sociologist Hadley Cantril analyzed the statistical results along with qualitative data from 135 interviews (Cantril 1966). Estimates based on the poll suggested that about 6 million Americans heard the broadcast; about 1.7 million mistakenly thought the realistic play was a news event, and about 1.2 million were frightened by it.

Unfortunately, respondents were not asked whether they believed in extraterrestrials nor encouraged to express their views on the desirability of contacting them. The focus of the research was on why some individuals apparently panicked. The most interesting explanation for the alleged panic focused on the transference of terrestrial war fears onto the Martians.

Public expression of fear over a specific danger may often be magnified if people have other worries that they can do nothing about. The fictitious Martian invasion came at a time when people all over the world were anxious about a real invasion. Hitler had begun his march toward war and was just about to seize a big chunk of Czechoslovakia. Fear of war had been constantly stimulated by all the papers. It is possible, therefore, that the Martian invasion gave people the opportunity to express their anxiety by focusing on a specific disaster. It could be called a *summary event*—a concrete representation of a vague but intense social and emotional situation (Rosengren, Arvidson, and Sturesson 1975).

In their issues for November 7, 1938, both *Time* and *Newsweek* made this point. *Newsweek* explained that on many people "already made danger-conscious by the recent war scare the effect of Wells-Welles realism was galvanic." *Time* was surprised that so many people had failed to understand that the program was pure fiction and concluded: "The only explanation for the badly panicked thousands—who evidently had neither given themselves the pleasure of familiarizing themselves with Wells's famous book nor had the wit to confirm or deny the catastrophe by dialing another station—is that recent concern over a possible European Armageddon has badly spooked the U.S. public."

People may project their desires as well as their fears onto extraterrestrials, as proven by a classic study carried out by Festinger, Riecken, and Schachter (1956). Pretending to be true believers rather than researchers, Festinger's team gained access to a small millenarian cult that claimed to be in mental contact with extraterrestrials. The cultists asserted that flying saucers would soon rescue them from a cataclysmic disaster that would annihilate their neighbors. The hostility of the cult toward nonmembers seemed to express resentment over low or ambiguous status in society, and the expected interplanetary salvation would exalt members far above ordinary people. Alison Lurie (1967) has written an engaging novel based on this study, and both books show how immodest desires and wishful thinking can combine to produce tragicomical results. But neither the 1938 poll nor this field observational study of the 1950s sought to determine the range of images of extraterrestrials held by members of modern society.

In recent years, major national polls have occasionally included a question on extraterrestrial intelligence, originally as part of human interest surveys about the flying saucer craze. In April, 1966, 34 percent of the 1,500 respondents to a Gallup poll said "yes" when asked, "Do you think there are people somewhat like ourselves living on other planets in the Universe?" Although 20 percent had no opinion, fully 46 percent said "no." By November 1973, the "yes" response had risen to 46 percent, and "no" had dropped to 38 (Gallup Opinion Index 1974).

Some of the same variables that predict support for the space program predict acceptance of extraterrestrial intelligence. Among college-educated respondents in 1973, 58 percent said "yes" compared to only 29 percent of those who had not even attended high school. Differences by annual income were similar, 57 percent of those whose families earned $20,000 per year believing in ETs, compared with 34 percent of those earning under $3,000. Only 30 percent of those without jobs believed, compared with 47 percent of manual workers, 52 percent of those in professional and business careers, and 58 percent of those in clerical and sales jobs.

Men were slightly more positive than women, 48 percent versus 45 percent, and Catholics slightly more than Protestants, 46 percent versus 43 percent. These differences may not be significant. Residents of rural areas, and of the Old South, were less likely than the average citizen to believe in ETs. The most potent variable of all was age. Fully 66 percent of respondents under age 25 said "yes," compared with only 31 percent of those 50 or over. Thus, acceptance of the possibility of extraterrestrial intelligence was a characteristic not only of the dominant classes in society, but also of the younger generation who will guide the course of civilization in the early decades of the next century.

Although something can be learned from occasional national poll questions, we must conduct fresh surveys to understand in any depth how American culture conceptualizes communication with extraterrestrials.

CONSEQUENCES OF CETI

In Spring 1983, when local publicity about CETI stimulated thinking on the campus (*Harvard Gazette* 1983), my seminar on sociological field research administered a brief, open-ended CETI questionnaire to 212 Harvard undergraduates. The questions were given through an interview that began with the following statement:

> Recently, there has been much discussion of the possibility that intelligent creatures might exist on other planets. A number of projects are under way or in the planning stages, designed to search the skies with sensitive radio receivers for any signals which might have been broadcast by civilizations far across the galaxy. For example, Harvard professor Paul Horowitz is now using the university's 84-foot radio telescope in a four-year search for evidence of intelligent life in the universe. We would like to know your thoughts and feelings about projects like this.

The first four questions explicitly sought opinions about the possible consequences of CETI: "What do you think the main consequence would be for people on our planet, if a research project did in fact succeed in picking up and understanding radio signals from other planets?" "In your opinion, what would be the *best* consequence which might result?" "What do you think would be the *worst* consequence which might result?" "Is there any other consequence which you, personally, *hope* might result?" The interviewers were instructed to write down exactly what the respondent said, although I am sure some simplification took place during hurried interviews. On file cards I copied each distinct idea about CETI from each questionnaire, a total of 1,082 cards.

The item ending the questionnaire asked for an evaluation: "Finally, we would like to know how you feel about the plan to search for radio signals from other planets. Please tell us whether you think this is a good idea or a bad idea, and give the main reason you have for your feelings about this." Comments made in response that seemed to discuss consequences filled another 191 cards.

Questionnaire S1986A included the first of these open-ended

items, and students also were asked to write comments after checking a box for the following fixed-choice item: "Do you think we should attempt to communicate with intelligent beings on other planets, perhaps using radio?"

The 1986 CETI responses were written on cards, bringing the total to 2,634. Rather than invest in a quantitative second-stage survey, based on items distilled from this vast collection of verbiage, I tried a different approach, using a panel of four judges to categorize the cards and generate quantitative data. The four were students in a special research seminar on spaceflight I taught at Harvard, two of them having already helped me administer S1986B when they were students in my survey research methods course. The procedure required each judge to sort the cards into categories he discerned in the data himself, and a special computer program combined these four independent classification systems.

I removed from the set any cards that only one judge could place in a definite category describing a consequence of CETI. This left a total of 2,109. Following a procedure I call *holotype analysis* (Bainbridge 1989b), the computer was able to discover 64 groups of from 2 to 262 cards, including all but 158 of the 2,109. Each group consists of a set of cards expressing its central idea, what I call the *holotype*, surrounded by others of similar meaning. For each one I wrote a final summary statement, based on the statements by the four judges, supplemented or clarified when necessary by material from cards in the holotype or its group.

As a practical matter, we shall focus on the twenty-eight groups consisting of fifteen or more cards. Often, smaller categories are muddled or redundant, because the judges could not agree on how to separate their cards from those in other groups. Table 8.1 lists my summary statements for the twenty-eight, a total of 1,757 cards.

It is gratifying to see a positive consequence, the gaining of knowledge, at the top of the list. World unity, in second place, also seems positive. But, in third place stands terrestrial panic, a negative consequence. Interplanetary war appears three times in the list: the fifth consequence has them attacking us; the twentieth has us attacking them, and the twenty-seventh suggests war could break out without a clear aggressor. If combined, the three about war would jump into second place, with 175 cards, 10.0 percent of the total. We must remember that the original statements were collected in a complex manner that often demanded very good or very bad consequences from respondents, and so we should not place too much confidence in such quantitative comparisons. Examination of the meaning of the twenty-eight ideas is more valid.

Table 8.1. Consequences of CETI in 1,757 Statements

Consequence of CETI	Cards in the Category Number	Percent
1. They would teach us much, giving us tremendous gains in knowledge.	262	14.9
2. The peoples of our world would be united in peace, either in common opposition to the extraterrestrials or in realization of our shared humanity.	173	9.8
3. The Earth would be gripped by fear, panic, hysteria, and paranoia.	166	9.4
4. Mankind would be humbled and forced to give up our arrogant belief that we are the center of the universe.	143	8.1
5. The aliens would make war upon us, to enslave or destroy us.	111	6.3
6. There would be a profound crisis in organized religion as people became disillusioned with traditional faiths.	86	4.9
7. We would gain great technological and scientific advances, some given us by the extraterrestrials and others learned through our own efforts.	63	3.6
8. We would greatly increase our investments in space research and exploration.	93	5.3
9. For most people, the consequences would be hardly noticed or quite insignificant.	55	3.1
10. Mutual exchange of culture would profoundly change the culture of Earth.	50	2.8
11. We would feel intense curiosity.	35	2.0
12. To learn that intelligent life exists on other planets would teach us we are not alone in the universe.	36	2.0

Table 8.1. *(continued)*

Consequence of CETI	Cards in the Category	
	Number	*Percent*
13. We would attempt to establish more complete communication with them.	55	3.1
14. They would help us cure diseases, such as cancer or AIDS, and perhaps even achieve eternal life.	58	3.3
15. We would establish a friendly alliance with the extraterrestrials.	24	1.4
16. Direct contact would be established, when we sent an expedition to their planet, or they came here to visit us.	27	1.5
17. Tremendous excitement would sweep the Earth.	27	1.5
18. It would broaden people's horizons, giving us brand new perspectives and possibilities.	44	2.5
19. There would be a boom in science fiction movies, TV shows, and books.	19	1.1
20. Earthlings would make war upon the aliens, to enslave or destroy them.	27	1.5
21. We would gain a greater understanding of how life evolved and of our own origins.	32	1.8
22. We would colonize planets, whether their world or others, reducing our problem of overpopulation.	16	0.9
23. We would reevaluate our philosophies, religions, world view, and ideas of the meaning of life.	34	1.9
24. It would distract us and divert funding from problems on our world which must be solved	16	0.9
25. With the extraterrestrials' help, we would transform our society to achieve a better, more peaceful way of life.	34	1.9
26. The extraterrestrials would introduce new ideas, new ways of thinking that would give us better perspectives and a more moral approach to life.	16	0.9
27. War would break out between us and the aliens, without either side necessarily being the aggressor.	37	2.1
28. People would doubt the report of messages from extraterrestrials, thinking it might be a hoax.	18	1.0

The specific kinds of knowledge to be gained, as expressed by utterances on cards collected in the first holotype, are both scientific and cultural, not just "technological, but living in harmony." "We might learn something useful" and thus "profit from another life form's knowledge," but we also will gain "knowledge for the sake of knowledge." "Imagine how much more there is to know—things that have more than human significance!" "We might learn amazing things," in an "explosion of knowledge never seen before." We might learn new "systems of rationality" and gain "mathematical advances." With "greater understanding of natural forces" we would "see if our laws of physics, chemistry, etc. are, indeed, 'universal.'" One Harvard student commented that "Two heads are better than one," presumably referring to the combined knowledge of two civilizations rather than to the possibility that the extraterrestrials are bicephalic.

Respondents contributing to the second category felt that CETI "would have a unifying effect on the planet. Dissentions here on Earth would seem petty or insignificant in comparison with relations with people from another planet." "It would make nationalism seem ridiculous, and we would end up with more global consciousness." "People on Earth would become more aware of their galactic community and begin to act in a more responsible, conscious, and just manner to their fellow humans." We might achieve "peace when our world draws together to define ourselves in relation to the other beings," and we "would realize that our petty battles are meaningless." Awareness of alien civilizations would make us "realize that as humans we have so much in common, and there is really not that much that has to divide us," "hopefully fostering a 'species-oriented' outlook rather than 'nation-oriented' or 'race-oriented'" outlook, with "pride in being an 'Earthling.'" "We would find that we are all humans—Russians, Iranians, Americans"—achieving "an end to racism and a new start for humanity. We're all one race; we can go to the stars together."

Cards in the third holotype spoke of "fear," "panic," "shock," and "mass hysteria" like the "reaction to *War of the Worlds*" "due to fear of invasion (H. G. Wells syndrome)." "Many people would be scared from seeing too many movies" about aliens. "Raw unadulterated terror" and "general pandemonium" would sweep the Earth. We would see the aliens as a "new enemy to be afraid of," "and God only knows how much more advanced they are than us." "Paranoia would develop over the idea that the other beings are superior and want to take over." Much of the panic would be "fear of the unknown." We would "worry about protecting ourselves before we even have any real notions of what they are about." "Defensive chaos" would ensue; "the military would want to nuke them." "A panic and mistrust of alien things" might develop into a general "xenophobia, paranoia giving rise to suppression of minorities" here on Earth. People "wouldn't know what to do or how to react," so we would drown in a "global panic, because of divergent interpretations and understandings." "Western civilization would be turned upside down," producing "total social upheaval, massive conflict between social groups." In the final analysis, "when belief systems are shattered, then everything falls apart." Thus, much of this "mass hysteria and anomie" would result from "people realizing that we are no longer the gems of the universe."

The fourth category sees the positive side of the radical self-reexamination that would be forced on humanity. It would "dethrone us," "teach us a little humility," "make us less conceited." It would destroy our "anthropocentrism," "grandiosity," and "egocentrism." "Realizing that there are other intelligent beings besides ourselves" would be a "lesson in modesty," and we would find "greater humility in the knowledge that we are not unique." "People would realize that Earth is a small part of the universe; perhaps we'd become more tolerant and less proud." "This would be a final blow to man's ideas of self importance and centrality in the universe." "It would be a very humbling experience because many people believe that there is no life on other planets and that we are very special 'chosen' beings!" "Obviously, we are no longer 'God's only children,'" and "people will become less inclined to think of ourselves as stepping stones to divinity." "Maybe we would have to reverse our supremist (I control all I touch) attitudes." At the worst, "we might get a cosmic inferiority complex, and our morale would slip downwards. We would feel insignificant next to these great, superior beings." At the best, this lesson would "help us to develop intellectually," "making us more responsible for our actions" and "more responsible to the environment" so "we wouldn't abuse Earth so much" with "a more balanced view of life in which man does not so arrogantly distinguish himself as separate from plants and 'lower' animals."

The three categories about war seem to assume that rapid travel between solar systems is feasible, although with a stretch of the imagination I can postulate forms of psychological warfare that would not require direct contact between humans and aliens. If the aliens were by nature "malicious," "violent," "unfriendly," or "meaner than us" they "would subjugate our people or destroy our Earth." One respondent acknowledged the possibility of "imperialistic aliens who do to us what advanced Western societies did to primitive societies—destroyed them," but he felt this was "not likely" because "imperialist society wouldn't have united" its entire planet, a necessity before it could war on other worlds.

Other respondents did not specify the type of society that might attack us, merely noting, "if they have an evil political credo, this could be a threat to our security." And the aliens might attack not only because they are warlike but because "they decide we're too bellicose." "Were we to reveal our location to a superior, warlike class of beings, and were they to analyze our history, we might suffer preemptive assault." Indeed, several students felt that Earth would strike the first blow. CETI would lead to our "development of military space power," "and we'd spend a lot of time and money finding and killing them." One expected "an *E.T.*-like response. If there were intelligent creatures, we would abuse them unthinkingly." Another saw it in terms of our colonial past. "We will probably exploit and ruin that life form, just as the Europeans 'killed' off the American Indians and their culture when they came to the New World." Others feared "becoming slaves or slavemasters, depending on who is more advanced."

Those who said CETI would cause a profound crisis in organized religion were split between opponents and defenders of the traditional faiths. One thought the best consequence of CETI would be the "serious blow to religion," whereas three thought the worst consequence would be: "skepticism about religion," "loss of faith in religion," and "tremendous stress on organized religion—what do Adam and Eve now mean?" Discovery of extraterrestrial civilizations would be "further evidence of Biblical folly" that might "smash religion." The results could be "severe demoralization of the very religious" and "the weakening of religious dogmas." Although some terrestrial creeds might be compatible with the existence of ETs, many "religions would have serious problems if the other beings had no concept of God."

Most of the remaining categories are self-explanatory, and their holotypes do not contain much detail or any noteworthy expressions. The tenth collection of cards stressed many aspects of possible cultural enrichment, including "new forms of music, arts, literature." Group 12, "an awareness that we're not alone," caused one of my judges to comment, "Most cards express the idea that man is 'lonely' now but will not be so when we find our brothers of the universe." As one card put it, CETI "would eliminate the loneliness of being solitary living particles floating through space on an infinitely turning sphere in an infinitely endless vastness of a vacuum." A card in group 16 predicted "we would send ambassadors, scientists, sociologists and many others to study their planet," and another expressed a personal desire "to meet one of them." One in group 17 showed the writer's excitement: "Inconceivable delight would invade everyone!"

The eighteenth group of cards, about broadening of horizons, resonates with categories described earlier that also postulated an intellectual revolution on Earth. CETI "would expand our horizons and possibilities in unprecedented ways," "widen our perspective, allow us to see ourselves more clearly," and "make people more open to new ideas." "It would stimulate people's imagination and awaken the dormant creativity within themselves." "Suddenly there'd be no end to our limits, our potential." CETI would mean "tremendous expansion of our horizons and the way we conceive of life," "broadened perspective by government and people," making all of us "look more at the meaning of life." "It will make people more interested in space research, which will expand their horizons beyond the pettiness of daily living." CETI may be a vain dream, but "it broadens people's perspectives and possibilities whether imagined or real. It is good to search for that which seems impossible."

Category 23 says much the same thing. "People would begin to think about what life really means. There would definitely be a philosophical reevaluation of the value of life." "The major consequence would be the change in people's attitudes toward life itself."

Category 25 suggests we might emulate extraterrestrial society. "Discovery of an alternate way of being," especially "if the planet were more advanced than our own," "could show us what is right or wrong with ours." "If these people had some sort of knowledge about how a civilization survives and how peaceful coexistence could be brought about," they would "allow us to see how to solve our problems by seeing how they've solved theirs." "Their life could be better; we could borrow and copy from them." "They might show us how to cope better with the power we have," "how to advance without threatening each other and the planet (example nukes)," how to "transform our society" "to be more peaceful" and achieve "effective management of our planet." We would learn how to be better human beings from learning from a higher form of life," getting "better (new) ideas for social organization" and discovering "a whole new way of being and living in which people coexisted peacefully and in community." "Contact would mean a new, and hopefully better era for mankind," with "increased respect and reverence for life."

The few cards in group 26 continue these meditations. CETI is "one of the most Earth-shattering things that could happen; disrupt all previous thinking and set us on the right track." First might come "shock—then an explosion of new ideas," "new perspectives on moral problems," a "revolution in human thought processes and patterns," a "renaissance in art, music, thought, and understanding."

In contrast to this theme of cosmic enlightenment, cards in category 24 say that investment in CETI and its follow-up projects would divert funding and concern from more important issues. "Money would be spent on research that would be sorely needed elsewhere," "for social programs" and "urgent needs." "This would distract us from more pressing problems on Earth. We would use it as an excuse to forget about our real problems." "We would use this as an escape from our own problems and not continue to solve major problems that face us here—politics, economics, class, ethnicity."

Finally, category 28 predicts many people would refuse to believe reports about communication with extraterrestrials. "Doubt if anyone would believe it. People are skeptics." "I think there would be a lot of denial from some segments of society—deny that the signals are produced by intelligent life." "There would probably be allegations of fraud, and there would be efforts to prove what happened." "People would think that it was a government conspiracy trying to get more money for its space program." "People in countries like Iran would be told by their leaders not to believe it."

FACTORS SHAPING CETI ATTITUDES

The S1981 data, based on the survey of 1,465 students at the University of Washington, can help us understand how acceptance of the existence of extraterrestrials varies across groups and identify the segments of the population most supportive of an attempt to communicate with them. We shall compare attitudes toward CETI with attitudes toward the general space program, thus tying this chapter to the others. We will find that factors which predict support for NASA funding also predict pro-CETI sentiments, but they do so much more weakly. Ideological factors we shall examine include opinions about the military, technology, various academic fields, and religion. We will introduce the four CETI items in the survey while considering sex differences in opinions about the subject.

Table 8.2 shows the patterns of response to two S1981 policy statements that advocated CETI and the search for life on other planets. Many of the questionnaire items were phrased as statements, and the respondent was asked to check one of five replies: "strongly agree," "agree," "neutral," "disagree," or "strongly disagree." Although about a third of the students are neutral toward each of the statements, it is encouraging to see that over 40 percent feel "We should attempt to communicate with intelligent beings on other planets, perhaps using radio." As also reported in Table 2.6, an equal number agree that, "Space exploration must continue so we can learn if there is life on other planets."

Table 8.2. University of Washington Responses to Two ET Policy Statements (percent)

	We should attempt to communicate with intelligent beings on other planets, perhaps using radio.			Space exploration must continue so we can learn if there is life on other planets.		
	792	639	1431	742	615	1357
	Women	Men	Total	Women	Men	Total
Strongly agree	6.1	13.6	9.4	5.9	10.7	8.1
Agree	28.0	37.9	32.4	37.2	33.5	35.5
Neutral	38.3	26.4	33.0	38.3	31.7	35.3
Disagree	21.0	16.9	19.1	16.7	20.3	18.3
Strongly disagree	6.7	5.2	6.0	1.9	3.7	2.7
	100	100	100	100	100	100

Opinion surveys often find a larger undecided vote among women than among men, and markedly greater percentages of the women responded "neutral" to our two ET policy statements. Whereas 38.3 percent of the women were neutral toward CETI, only 26.4 percent of the men could not take a stand pro or con, and men give greater support to the attempt to communicate than women. The second item in Table 8.2, about continuing space exploration to find life, does not really show greater male enthusiasm. Rather, men are more polarized than women, showing greater negative as well as positive responses.

The two other CETI items in S1981 ask for opinions about facts, yet they, too, measure attitudes toward the subject. One was drawn from S1977 and summarized what many people feel might be the cultural benefits of CETI: "Communication with intelligent beings from other planets would give us completely new perceptions of humanity, new art, philosophy and science." As Table 2.6 reported, 61.6 percent of the respondents agreed. The final item was phrased negatively, so I could be confident my findings would not merely be the result of positive response biases, and it refers to the assumption underlying CETI: "Intelligent life probably does not exist on any planet but our own." Overwhelmingly, university students rejected this statement, only 8.4 percent of the men and 12.0 percent of the women checking the box for "agree" or "strongly agree." However, distribution across the five response categories was not so badly skewed as to compromise correlational analysis.

I constructed a *CETI-scale* out of these four statements, arriving at a more general measure of attitudes than any one item taken alone. This scale awards a respondent 1 point for each pro-CETI attitude—that is, for agreeing to some extent with each of the first three CETI statements and for disagreeing with the fourth. Respondents who skipped any one of the four statements were dropped from consideration, and a total of 1,350 both have a valid score on this CETI-scale and told us their gender. With the sexes combined, 15.0 percent have a score of 0, expressing pure anti-CETI opinions. The distribution is quite even, with 18.9 percent having a score of 1, 22.7 percent a score of 2, 21.3 percent a score of 3, and 22.1 percent a score of 4, which represents the extreme in pro-CETI sentiments. Here, again, the men are slightly more pro-CETI than the women.

In previous chapters, we generally relied on Pearson's r to measure correlations, but the extensive use of ordinal variables in this chapter requires us to switch to tau, which is technically more appropriate. In so doing, we must adjust our standards for judging the magnitude of coefficients, because tau will give us numbers much closer to zero, even when the association is extremely strong. For example, the tau of 0.16 between being male and score on the CETI-scale is quite respectable and achieves the 0.0001 level of statistical significance.

With the individual items and the scale created from them, we can examine the connection between CETI attitudes and various social and ideological factors, starting with two antispace items and two pro-space items. Altogether, only 17.6 percent of the university students agreed that, "The United States is spending too much money on space, so appropriations for the space program should be reduced." A minority of the students, 22.6 percent, felt that, "Space exploration should be delayed until we have solved more of our problems here on Earth." And an absolute majority, 52.7 percent, believed, "In the long run, discoveries in our space program will have a big payoff for the average person." Support for NASA funding was also high. The CETI scale achieves respectable correlations with these items (tau = -0.34 , -0.29, 0.31 and 0.35).

Both CETI and spaceflight depend on advanced technology, and a counterculture of antitechnology ("Luddite") sentiments currently erodes public support for progress as it has been known in recent centuries. S1981 included five antitechnology items, all phrased as agree-disagree statements: "Machines have thrown too many people out of work." "It would be nice if we would stop building so many factories and go back to nature." "Technology has made life too complicated." "All nuclear power plants should be shut down or converted to safer fuels." "People today have become too dependent upon machines." On average, these items achieve a tau of -0.18 with support for increased space funding, but only -0.08 with supporting an attempt at communication with extraterrestrials and -0.09 with the more sensitive CETI scale.

Thus, respondents who agree with these indictments of modern technology and of industrial society show a slight tendency to oppose CETI, although their opposition to increased funding for the space program is greater. The weak associations suggest there is both need and opportunity for stressing the open, humane, nontechnical qualities of CETI. This has been done in some very effective pro-CETI propaganda, the movies *Close*

Encounters and *E.T.* The first official close encounter between extraterrestrials and American bureaucracy takes place via the humane medium of music. When the stranded E.T. assembles a CETI contraption out of common household junk in order to "phone home," he places CETI in a familiar, humanistic context. Interstellar contact and communication become expressions of friendship, in these films, rather than being the cold dreams of a high-tech elite. Thus, although the search for extraterrestrial civilizations does require the use of advanced technology, it may be possible to present it to the public as a triumph of the human spirit, rather than as an arcane technical breakthrough that might antagonize technophobic segments of the population.

Previous research has confirmed the commonsensical proposition that science fiction encourages pro-CETI attitudes (Bainbridge, 1976, 1982a, 1986), even examples of this popular genre emphasizing mystical rather than technical values. One theme in both *Close Encounters* and *E.T.* is that ET contact is for the ordinary people of the world, not just for technical and political elites, and the people loved these movies. Unlike the Space Shuttle, where only experts may go, the messages possibly beamed toward us from other stars can be shared by all the citizens of Earth.

The fact that attitudes toward technology play a role, albeit an exceedingly minor one, in determining attitudes toward CETI suggests we ought to look for other measures of intellectual value-commitment. Because our respondents are university students, the most salient ideological orientations are probably expressed through their preferences for college subjects. Questionnaire S1981 included a series of thirty-two items asking students how much they like particular fields of study, using a 7-point preference scale, and I examined their correlations with the CETI scale.

Appreciation of Astronomy predicts support for the space program and for CETI, but again the correlation for space funding (tau = 0.26) is stronger than for supporting communication (0.19) or the CETI scale (0.23). For Physics and Engineering the associations with space funding (0.22 and 0.20) are much stronger than those with the CETI scale (0.10 and 0.08), and for

Mathematics the CETI associations vanish. The strength of Astronomy suggests that support for CETI is encouraged specifically by curiosity about the universe around us and only weakly if at all by a positive orientation toward the physical sciences as such.

Neither increased space program funding nor the CETI scale correlate with preferences for three life sciences: botany, biology, and zoology. Apparently, interest in the miracle of life does not automatically produce a desire to seek out life in all the forms in which it is manifested, off the Earth as well as on. Nor are there connections linking the CETI scale to feelings about three subjects stressing communication between people: communications, foreign languages, and sociology. These social fields encourage weak opposition to space program funding, with taus averaging -0.09. Communications students narrowly focus on communication with members of their own society and their department is a training ground for professionals in broadcasting and the press. Sociologists study Americans, and even students of foreign languages generally reach only as far as European cultures.

Anthropology achieves a significant tau of 0.14 with the CETI scale, a weak coefficient even for tau, but worth considering. Apparently, people who love alien terrestrial cultures are prepared to extend their xenophilia beyond the Earth. And Anthropology is not associated with general support for the space program. In contrast, students who like Social Work, tend not to favor increased space funding (tau = -0.19) and rate CETI scale items slightly lower than do others (-0.08).

The academic preference questions included two broader items, "the sciences in general" and "the humanities in general." The sciences support the space program (0.18) whereas the humanities oppose it (-0.11). These coefficients are significant at the 0.0001 level, but they are not large, even for taus. This finding would have been predicted by C. P. Snow's theory that modern intellectuals are divided into two opposed cultures, the sciences and humanities, with contradictory values (Snow 1969). Interest in the sciences aids CETI (0.12), whereas the humanities are neutral to it (0.03).

Thus, except for Social Work, we do not see sources of opposition in the value systems expressed through scientific and scholarly fields. We find support in Astronomy and Anthropology. If Social Work represents opposition, we may hypothesize that the political left rejects CETI, but other analyses suggest weak opposition from the political right. CETI appears ideological poles apart from the military applications of space, and the GSS political question reveals a very weak negative correlation (tau = -0.11, significant at the 0.003 level) between conservatism and the CETI scale. One possible explanation for this complexity may be found in conservative religious faith.

RELIGIOUS OPPOSITION TO CETI

Although the possibility of contact with extraterrestrial intelligence derives from recent developments in science and technology, it reminds us of age-old questions of a philosophical or religious nature. Where have we come from? Who are we? Where are we going? What is Man in relation to the cosmos? Are there forms of consciousness vastly superior to human beings? What hope can there be for the distant future? Traditional Christian religion provided answers for all these questions, answers that have been shaken already by discoveries in the sciences and might be shattered completely by contact with extraterrestrial civilizations. Thus, CETI serves motives that border on the religious, yet it threatens particular religious doctrines and values that have been important in the Christian tradition.

Questionnaire S1981 contained several well-tested questions on religion, permitting us to look deeply into the relationship between traditional religious faith and support for CETI. Of the 597 respondents who could be identified as Protestants, 38.0 percent felt we should attempt to communicate with intelligent beings on other planets. This contrasts with 41.8 percent of the 55 Jews, 43.8 percent of the 368 Catholics, and 50.3 percent of the 266 who said they had no religion. Clearly, something in Protestantism reduces support, and perhaps something in irreligiousness encourages it. A difference of 12.3 percentage points is not to be ignored, so in Table 8.3 we examine the religion effects further.

Table 8.3. Religion and Support for CETI

Religiosity	Increase Space Program Funding	We Should Attempt CETI	CETI Would Give Us...	Learn If Life Is on Planets	ETI Does Not Exist	CETI Scale
All Respondents:						
Being "born again"	-0.10*	-0.16*	-0.16*	-0.13*	0.16*	-0.21*
Belief in miracles	-0.13*	-0.17*	-0.15*	-0.09*	0.18*	-0.18*
Beliefs are important	-0.14*	-0.16*	-0.14*	-0.11*	0.15*	-0.19*
Church attendance	-0.12*	-0.16*	-0.16*	-0.08*	0.18*	-0.18*
Male Protestants:						
Being "born again"	-0.15	-0.37*	-0.31*	-0.32*	0.35*	-0.45*
Belief in miracles	-0.13	-0.33*	-0.26*	-0.21*	0.33*	-0.37*
Beliefs are important	-0.14	-0.30*	-0.19*	-0.24*	0.27*	-0.35*
Church attendance	-0.11	-0.27*	-0.22*	-0.21*	0.30*	-0.34*
Female Protestants:						
Being "born again"	-0.11	-0.14	-0.18	-0.10	0.19	-0.19
Belief in miracles	-0.18*	-0.17*	-0.15	-0.11	0.19*	-0.19*
Beliefs are important	-0.12	-0.19*	-0.16	-0.12	0.18*	-0.20*
Church attendance	-0.11	-0.14	-0.14	-0.10	0.17*	-0.17*

Correlation (tau) with

*Significant at the 0.0001 level.

Several questions on religion gave very much the same results, and I chose four representing different forms of involvement in the sacred. The first was drawn from the Gallup Poll: "Would you say that you have been 'born again' or have had a 'born again' experience—that is, a turning point in your life when you committed yourself to Christ?" Of the 1,421 respondents who answered this question, 22.2 percent admitted to being Born Again Christians, compared with 34 percent nationally (Gallup Opinion Index 1977). The second item was one of my standard agree-disagree statements: "Miracles actually happened just as the Bible says they did." Fully 46.8 percent of the respondents gave some measure of assent to this fundamentalist opinion. The third item asked, "How important to you are your religious beliefs?" Respondents checked one of four boxes: "very important" (32.5 percent), "fairly important" (33.4 percent), "not too important" (22.6 percent), and "not at all important" (11.5 percent). The final religion item inquired how often the respondent attended religious services, on a 6-point scale from "once a week or more" to "never." Although 16.0 percent never attended church, 41.5 percent did so at least once a month.

The first part of Table 8.3 looks at correlations based on all respondents to the survey, and here we do indeed see significant associations. Religious students are less likely to support increased funding for the space program, and are perhaps even less likely to support CETI. Indeed, the consistent negative correlations between the religion items and the CETI-scale indicate that traditional religion is a real force working against this great project to discover other civilizations in the universe.

I have been using tau to express the associations, because it provides the fairest comparisons across the variety of questionnaire items considered here. But, as mentioned earlier, tau generally produces smaller numbers than r, and it is hard to get a sense of the magnitude of religion's influence from these coefficients. Therefore, we should look at the percentages for one of the religion items. The four religion variables correlate highly with each other, and the easiest one to discuss, because it has only two categories, is being born again.

Whereas 22.8 percent of the born agains wanted space program expenditures increased, 32.2 percent of those not born again felt this way. Only 27.7 percent of the born agains wanted us to attempt CETI, compared with 46.0 percent of those not born again. Whereas 32.3 percent of the born agains supported the search for life on other planets, 47.1 percent of other respondents did. The difference was exactly 20 percentage points on expecting cultural benefits from CETI, 46.4 percent compared with 66.4 percent. Of born agains, 53.3 percent think it possible that intelligent life exists on other planets, disagreeing with the statement that they do not, compared with 72.2 percent of other respondents.

The born again phenomenon is a religious movement primarily within the Protestant tradition, often called *Evangelical Protestantism*. The fact that the denominational groups differed in their average level of support for CETI required that I run the religion computations for each group separately. The most interesting findings were for Protestants, the group with the lowest level of support, so in the middle and bottom of Table 8.3 we repeat the correlations just for them. Because gender is also a prime factor for religion as well as for space attitudes (Bainbridge and Hatch 1982), here I separate the men and women. Correlations for the Protestant women are almost identical to those for the entire set of respondents, whereas those for Protestant men are vastly stronger. Indeed, the male tau of -0.45 between being born again and the CETI-scale is majestic, about as big a tau as we would expect to find in questionnaire opinion data.

Previous research found that the Born Again Movement was the most powerful religious force on the university campus (Bainbridge and Stark 1981b). Being born again has a decisive impact. Although 33 percent of the Protestant born again men have a CETI-scale score of 0 and only 10 percent have a perfect pro-CETI score of 4, the figures are almost exactly reversed for Protestant men who are not born again. Only 6 percent of them have a CETI-scale score of 0 and 35 percent have a score of 4. For all Protestants, the born agains are half as likely as ordinary respondents to want us to attempt CETI, 23.3 percent versus 46.1 percent. Apparently for American Protestants, CETI is an issue of great potential religious significance.

Computation of the same correlations for Catholics failed to reveal a strong religious effect. That is, there was very little difference in support for CETI among Catholics who were very religious compared with those who were hardly religious at all. Also, there was little sex difference in the impact of the very weak religious effect that did appear. Clearly the Evangelical Born Again Movement pulls the Protestant support below that for other groups.

The sex difference among Protestants is not easy to explain. My best guess is that it has something to do with the fact that women are far more likely than men to give neutral responses to the questions. Men are more ready to express extreme opinions on CETI, and thus they may find it easier to use CETI as a medium for expressions of religious views. Religion normally is more salient for women than for men, and space projects are more salient for men.

For the 334 female Protestants who answered the questions, only 21.6 percent of the born agains support the attempt to achieve CETI, compared with 35.4 percent of those not born again. Among 250 male Protestants, 25.0 percent of the born agains support CETI, compared with 59.7 percent of those not born again. Thus, for Protestants, being male does increase support for CETI, whereas being born again greatly reduces it. The religion effect is so powerful that it deserves not only the notice of the aerospace community but also the attention of future sociological research projects.

One possible explanation for fundamentalist religious opposition to CETI is that the existence of extraterrestrial life and civilization would tend to refute Biblical stories of the origin of the Earth and its people. If so, then we would expect a strong connection between opposition to CETI and opposition to Darwinian theories of evolution. In fact, there is a significant negative correlation (tau = -0.22 for all respondents) between the CETI-scale and feeling that "Darwin's theory of evolution could not possibly be true." Unfortunately, close inspection of the data do not allow me to conclude that religion undercuts support for CETI specifically because it promulgates opposition to this scientific theory that harmonizes with the possibility of extraterrestrial life. Darwinian questionnaire items simply measure religious traditionalism (Bainbridge and Stark 1981a), and they do not reflect a cognitively independent attitude. Statistically, the Darwinian item is so highly correlated with straight religion items that it cannot be manipulated independent of them. The fact that favoring biological science was not connected to support for a CETI project testifies to the weakness of any explanation that makes the theory of evolution an effective intellectual link between fundamentalism and attitudes toward CETI.

Another approach is to note the large body of literature suggesting that sectarian, conservative Protestants (but not necessarily Catholics) exhibit a general xenophobia, a hatred and fear of outsiders. For example, one famous study found that rejection of Jews was especially common in fundamentalist Protestant groups, among the same population that opposes CETI (Glock and Stark 1966). The reasons for this xenophobia may be theological, as the original study said, or may stem from social and economic deprivations, as more recent commentators have suggested (Lipset and Raab 1970; Stark and Bainbridge 1985; cf. Gusfield 1963; Hofstadter 1963).

VOICES FROM THE SKY

We have just seen that the CETI movement is opposed implicitly by another modern social movement of great force: Protestant Evangelicalism. Perhaps, propagandizers for CETI can find a way to couch their hopes in terms more congenial to traditional religion, perhaps on the model of C. S. Lewis (1965; cf. Stevens 1969). But if social deprivations and cultural xenophobia are behind the evangelical opposition, there may be little that mere words can accomplish. The possibility that new religions just being born may be more supportive of exploration of the cosmos cannot affect things on the historically short run, and thus near-term CETI projects cannot expect much aid from this exotic direction (Bainbridge 1982b).

CETI gains strength from the general space program and thus should be furthered by its successes. But CETI gains also because it is not tied equally to all aspects of spaceflight. Furthermore, the fact that social factors predicting support for the space program are markedly weaker predictors of CETI attitudes suggests that the extraterrestrial contact movement can grow far beyond the limits of NASA's political constituency. Whereas only 30.2 percent of the S1981 respondents felt funding for the space exploration program should be increased, 41.7 percent want a new program to achieve CETI. Indeed, all our results indicate that enthusiasm for the search for extraterrestrial civilization is a powerful, independent idea capable of achieving a strong positive consensus among educated young people.

This growing hope that we may find brothers or mentors beyond the stars shares much in common with the Idealistic goals discussed in Chapter 5, and yet it is far more popular. Perhaps, ideals are more attractive and persuasive when they are personified, when their abstractness is rendered concrete through incarnation in more-or-less human form. From this perspective, Jesus is mercy personified. ETs may be imagined as saviors, teachers, friends, or demons. Spaceflight is a human adventure, and the limitations we perceive in human nature may limit our hopes that exploration of the universe will actually achieve valued ideals. CETI builds a relationship between humans and extrahumans who may take on superhuman attributes in our imaginations. And over the ages humans have poured their unrealized hopes into religions that listened for guidance from the heavens.

CHAPTER NINE

EPILOGUE

In 1985 I was asked to testify before the National Commission on Space, but I declined the invitation. The world does not need my personal opinions to guide it toward the best future. The practical benefits of spaceflight are well documented, and I had no right to pretend that my personal feelings expressed the idealism of American culture. To testify in the role of a social scientist, one needs data and theoretical findings, not mere intuitions and private enthusiasm. Having completed an extensive survey of our culture's goals in space, I am finally ready to testify. This book is my contribution to the public debate on space policy.

On the basis of thousands of questionnaires, administered as part of a systematic survey of the goals of space, I now can say that American culture posses a complex conception of the value of the space program. Whether we focus on more than a hundred specific ideas or on a few large classes of ideas, we find a rich set of purposes spaceflight may serve. From these data, scholars may learn much about our society's orientation toward science and technology. Members of the Spaceflight Movement can draw inspiration and an articulate expression of their creed from the words of the respondents and the mathematical structure of the correlations.

My own perspective is summarized most briefly in terms of two statements: one an expression of personal commitment, and the other a report of cool analysis. First, humanity should, at almost any cost, expand into space to become an interstellar civilization. Second, human beings will never colonize space as the result of rational analysis of the greatest benefit to individuals or to terrestrial society. This seems a case of you cannot get there from here.

Over the decade after this analysis was first published (Bainbridge 1976), I often worked in the sociology of religion, attempting to understand the social origins and patterns of the most noneconomic, apparently nonrational variety of human behavior. However, my approach was to postulate that religion, like more mundane features of human life, was an expression of human desires and a social institution through which people sought palpable rewards. Thus, my aim was to see if religion could be linked to motives and actions like those of economic behavior, to create a scientific bridge between the transcendent and the utilitarian. My success in that work will have to be judged by specialists in the sociology of religion or by you if you care to read my books in that field (Bainbridge 1978; Stark and Bainbridge 1985, 1987). But from my work on religion I gained fresh optimism that vast human energies could be harnessed to a transcendent task like interplanetary colonization.

Sociologists are commonly accused of letting their personal values distort their research, and a scholarly debate has raged for years concerning whether sociology could be value free. I subscribe to the moderate view that sociologists are entitled to select their research topics on the basis of their personal value commitments, but they should then study them in as scientific and objective a manner as they can.

For example, the overwhelming majority of sociologists who study race relations are strongly opposed to racism, and their research is intended to overcome this scourge. Although racists may resent the legions of social scientists arrayed against them, the essentially unanimous opposition of sociologists to racism does not necessarily mean that they do bad science when studying it. Indeed, they might hurt their cause rather than help it, if they committed scientific errors, whether intentional or unintentional. If we can learn conclusively the causes of racism and identify factors that will dissolve them, then the purpose to which so many sociologists have dedicated their lives will be advanced. Only when value commitments encourage wishful thinking or when prejudice toward political opponents causes a critic to misrepresent them unjustly, do values distort social science.

My research on spaceflight has been energized by the intense desire to see human civilization expand to the planets and the distant stars. In our present society, there is no consensus about spaceflight, as there may be about race relations, and I fully understand that many people disagree with me. Charitably, I think they simply place interstellar civilization far lower on their list of priorities than I do, and they may not have seriously informed themselves about the vast possibilities that exist beyond the mundane world with which they are familiar.

So, I have been gripped by a dream. But I have continually counseled myself to think clearly about the social and cultural reality from which any realization of that dream must arise. In particular, I have been struck by the uncertain message I read in the history of spaceflight and in the national polls.

However a hypothetical civilization might achieve spaceflight, ours did so as an arbitrary consequence of international military competition. The scientists did not successfully agitate for rockets to loft their instruments above the atmosphere; the philosophers did not conclude that human transcendence demanded interplanetary probes; the businessmen did not float into space on issues of interplanetary stocks and bonds; the general public did not rise up with one voice and shout: "Upward!" Instead, a coalition of fanatic spaceflight enthusiasts, beleaguered military agencies, and capricious political leaders focused enough effort to develop long range missiles using a technology that could be adapted to modest space projects. Whatever emphasis we place on each of these actors, the whole process was neither pretty nor inevitable.

Some proponents of spaceflight comfort themselves with the thought that ours is an aggressive, exploratory species, driven by instinct to conquer the universe. Of course, such a theory rests on very crude notions of "human instinct" and on questionable assumptions about how our species will express its innate drives in future eras. Behavioral scientists are not at all convinced that aggressive and exploratory instincts explain human history, and one never sees these notions in contemporary sociology. Instead, the standard social scientific theories hold that socio-economic factors produce large-scale collective action and cultural change, building on only the most general urges that might otherwise find expression through many different patterns of behavior, dependent upon conditions in the social environment.

Animate life is motivated to solve basic problems of food, security, and reproduction. Intelligence emerges in biological evolution only because it improves organisms' capacity to solve these problems. A young, intelligent species, like our own, retains many socially and biologically conditioned behavioral tendencies (customs and instincts, if you like) that helped our species solve these basic problems under the conditions in which it came into being. Now, technological development has transformed the conditions of life and presents us with many alternate means of satisfying our primary needs. Customs and instincts are fast becoming dysfunctional atavisms, to be discarded in the service of our need for security.

We have the capacity to provide food and shelter to all members of our species. The erotic urges that sustained human reproduction when high death rates demanded high birth rates have become disturbing influences rather than essential to our survival. But now they can be satisfied in many ways, and a worldwide conviction has emerged that unrestrained reproduction is no longer acceptable. Not only birth control techniques and formerly deviant sexual practices, but also the arts and many other forms of sublimation can satisfy these erotic urges. As I have noted, some people argue that interplanetary colonization could enable human population to expand indefinitely. But even the most rudimentary consideration of the mathematics of population growth proves that colonization could handle only an infinitesimal portion of the population increment that unfettered reproduction would produce. Thus, an "instinct" must be transformed, rather than letting it drive us into the universe.

I think it very likely that ancient intelligent species will have evolved static societies which have achieved containment of instinct—which have transformed themselves culturally and biologically to satisfy natural desires in the most efficient and safest ways. The social conditions that magnify aggressive and exploratory drives are highly dangerous. They generate pressures toward war before they stimulate interstellar colonization. Perhaps such "outward urges" could fuel an ambitious space program. But for individuals and small groups, quicker rewards of power and property can be obtained through competition against other individuals and groups. Thus, all civilizations that continue to be aggressive and expansionist will be politically unstable. When they reach our level of technological development, they enter a period of extreme danger, and the risk of nuclear annihilation is only the most obvious of the ways they could bring doom upon themselves.

At the same level of technological development, intelligent species acquire effective techniques for modifying themselves, culturally and biologically. Electronic communication and rapid transportation have made possible a stifling world government. Techniques such as genetic engineering, psychoactive drugs, and electronic control of the brain make possible a transformation of the species into docile, obedient, harmless organisms. Less dramatic constriction of social structures can achieve the same thing. Not interstellar flight but stasis becomes the order of the day—the policy of the millennium and of the eon. Some species may fail to transform themselves, and they will survive only briefly before destroying themselves in nuclear war or in some other suicidal catastrophe that we may not even yet imagine.

If interstellar colonization is ever to be possible, it must begin very rapidly, within a few short decades of the development of nuclear physics and biological engineering. Ordinary socioeconomic forces will be insufficient to launch galactic exploration this rapidly, and only transcendent social movements could possibly channel enough of a society's resources into the project to succeed before stasis or annihilation. Such a movement was able to exploit political and military tensions to achieve the first great steps in space, but entirely new social forces will be required to impel our species much further.

The national opinion polls provide only modest encouragement. Some of my fellow space enthusiasts like to note that a majority of respondents to many of these polls want space funding either increased or kept the same. The pessimist might note that a larger majority generally want space funding kept the same or reduced. True, the young and the educated are more enthusiastic than the old and the uneducated, and the societal elites of the future may steer an indifferent majority toward the planets. But the chief impetus remains with a minority.

This is one case where there is nothing wrong with minority rule. All that the Spaceflight Movement needs is a relatively modest investment of wealth and talent, perhaps ten times the present rate of funding. The indifferent majority can do what it wants. If the gross world product can be kept expanding, then only a small share will be sufficient for colonization of Mars and several moons over the course of a century. One does not have to be a Puritan to be concerned that America spends far more on alcohol or drugs or gambling than it does on the universe above our heads. If an organized minority demands that its share of taxes go toward exploration and habitation of the universe, then the indifferent majority and the minorities committed to other causes have no right to complain. At present, our civilization seems committed to accomplishing nothing.

I write these lines a few hours after the Galileo space probe flew past the planet Venus to get a gravity assist on its way to Jupiter. Nothing could be more symbolic of the confused state of the space program! Galileo was supposed to be launched directly to Jupiter in May 1986, taken to Earth orbit by the Challenger, and fired to its goal by a hot rocket carried in the shuttle's cargo bay. When the Challenger exploded, concerns about a possible second disaster substituted a less energetic upper-stage rocket, incapable of a direct flight. Over the next five years, Galileo is supposed to loop past the Earth twice, drift through the asteroid belt, and maybe go into orbit around Jupiter. The reader may already know whether this roundabout attempt was destined to succeed or fail. Even if it works, this is no way to explore the planets.

Consider a space analogy. The Saturn V was able to send

three men to the Moon in a few days, with 7.5 million pounds of thrust. Imagine a budgetcutter demanding that the Saturn V be built with only 1 million pounds of thrust, under the assumption that the trip easily can be lengthened to a few weeks. In truth, it would never get off the launch pad with a thrust that low, and even the lessened investment would be utterly wasted. If we do not move rapidly into space, we will not move at all.

NASA's plans for the next few decades are ambitious only in comparison to the inaction of the past twenty years. I was born only three years before the first flight of von Braun's V-2, yet the first human expedition to Mars is tentatively slated for about the year I am scheduled to die of old age. Speaking personally, that is too slow. But, if we conceptualize the present time as a period of gradual development sandwiched between periods of rapid activity, like the slow movement of a symphony, then there is room for optimism. And the results of my research contribute empirical substance to that breath of hope.

The high level of acceptance of Technical space goals assures that a modest near-Earth space program will continue, gradually developing the technology and the infrastructure that is required for major new expansion. Controversies over the Military goals may hurt public appreciation of spaceflight, but only to the extent that NASA permits itself to be seen as the handmaiden of the Defense Department, and some military applications incidentally increase the nation's spacefaring capacity. Although the Idealistic goals are rather low in popularity, they are the sort of motives that can create social movements that promote their causes aggressively, without the necessity of capturing the minds of the majority of citizens. If both Technical and Idealistic goals receive sufficient support, then the goals of the Colonization class may become serious projects for future collective achievement.

Whether out of rational analysis or wild dreaming, members of American culture have great hopes for communication with extraterrestrial beings. Without discrediting the utilitarian justifications for the space program, advocates for spaceflight should exploit this reservoir of social energy. Citizens of modern society know intellectually that the stars are not mere lanterns hung from the ceiling of the sky. But emotionally, most probably feel that nothing valuable exists beyond the boundaries of their own world. However vain its assumptions may turn out to be, the CETI craze at least awakens people to the fact that a vast and interesting universe spreads out to infinity above the clouds. That consciousness may allow Earthlings to realize the value of the vast range of goals we can seek in space.

The major classes of space goals reflect the general values that humans possess. The Technical goals express everyday desires for economic gain, improvement in our technological capability, and the increase of practical knowledge. The present moment shines with the promise of an end to war and the dissolving of Military motives into international cooperation, although certainly the world has known disappointments before. Idealistic goals resonate with a variety of human emotions and harmonize with movements toward social progress across a wide compass. Colonization would mean nothing less than a fresh start, not only for human society but for life itself, to be reborn endlessly on distant worlds.

The search for extraterrestrial life is most unpredictable in its outcome. All bets are off, if we actually receive a message from beyond this solar system. But even a vain belief in extraterrestrials may transform history. Perhaps a new religious denomination will appear, marching to the faith that the gods dwell somewhere across the universe, waiting for us to find them. Or perhaps the hope that will focus our energies skyward will be a pathetic feeling that other civilizations have solved the problems that threaten to destroy us, and they will give us guidance if only we can contact them. Stranger things have happened.

The first phase of space progress was achieved by a social movement operating outside the ordinary institutions of society, but exploiting them whenever possible. Future revolutionary progress may follow the same course. In the end, the earthbound governments that currently set modest space policies may have to be transcended or abandoned. At present, the Spaceflight Movement is biding its time, infusing the culture with its ideology, waiting for those cataclysmic social conditions that might motivate a new rush forward.

APPENDIX A: SPACE GOALS FROM THE 1977 SEATTLE VOTER SURVEY

These forty-nine space goals were rated by 225 Seattle voters, as explained in Chapter 2. For each, they checked a box to indicate how good a reason they thought it was for continuing the space program: not a good reason, slightly good reason, moderately good reason, and extremely good reason. For the mean ratings in the first column of the table, these four responses were coded from 0 (not a good reason) to 3 (extremely good reason). The second column shows the percent of those responding who described the goal as an extremely good reason for continuing the space program; depending on the item, from 215 to 225 voters responded. The third and fourth columns show the strongest correlation possessed by each goal with another and the number of that other goal in the list. The numbering of items is arbitrary, simply the arrangement that was used in the computer analysis, and the space goals were presented in five different, random orders in five editions of the actual questionnaire.

	Mean Rating	Percent High Rating	Strongest Link Correlation	Strongest Link With Item
1. Space has military applications; our nation must develop space weapons for its own defense.	1.34	19.6	0.64	27
2. Military reconnaissance satellites (spy satellites) further the cause of peace by	1.74	33.2	0.52	1

	Mean Rating	Percent High Rating	Strongest Link	
			Corre-lation	With Item
making secret preparations for war and sneak attacks almost impossible.				
3. Radio, telephone, and TV relay satellites are vital links in the world's communications system, fostering education and international understanding.	2.50	61.8	0.52	6
4. Earth resource satellites allow us to monitor the natural environment of the Earth and help locate valuable resources such as minerals and water.	2.25	45.0	0.52	15
5. Navigation satellites are a great help to ship and plane navigators, and traffic control from space can aid safe and efficient use of conventional transportation systems.	2.06	36.8	0.58	6
6. Meteorology satellites aid in making accurate predictions of the weather	2.30	51.4	0.58	5
7. Electric power generated in space and sent down to Earth will help solve the energy crisis without polluting our environment.	1.83	35.9	0.46	9
8. Raw materials from the moon and other planets can supplement the dwindling natural resources of the Earth.	1.47	25.2	0.58	32
9. Commercial manufacturing can be done in space without polluting the Earth; completely new materials and products can be made in space.	1.19	15.2	0.59	22
10. Outer space will be used in the disposal of very dangerous waste	0.89	13.2	0.41	22

	Mean Rating	Percent High Rating	Strongest Link	
			Corre-lation	With Item
products, such as unwanted radioactive materials.				
11. Space hospitals put into orbit where there is no gravity will be able to provide new kinds of medical treatment and give many patients easier recoveries.	1.42	17.7	0.55	15
12. Rockets developed for spaceflight will be used for very rapid transportation of people, military equipment, or commercial goods over long distances on the Earth.	1.31	14.1	0.54	15
13. Space technology produces many valuable inventions and discoveries which have unexpected applications in industry or everyday life.	2.25	48.2	0.64	21
14. Space development will give us new practical knowledge that can be used to improve human life.	2.12	42.4	0.65	21
15. Space technology will allow us to manage the environment of our planet because it is developing techniques for managing artificial environments that support human life.	1.58	20.4	0.55	11
16. We must continue the space program in order to maintain the quality of American technology.	1.52	20.6	0.62	24
17. The space program provides an essential stimulus to the whole economy by investing money and paying employees.	1.24	10.9	0.60	29
18. Space exploration is a positive substitute for war because it channels man's aggressive	1.28	16.1	0.47	35

	Mean Rating	Percent High Rating	Strongest Link	
			Corre-lation	With Item
instincts into non-military activities.				
19. We must go beyond the finite Earth into infinite space in order to continue economic growth without limit.	0.69	6.5	0.56	32
20. Overpopulation on Earth can be solved by using the living space on other planets.	0.79	11.3	0.61	32
21. Space exploration adds tremendously to our scientific knowledge.	2.35	52.7	0.65	14
22. We can conduct certain dangerous kinds of scientific experiment far in space so accidents and other hazards will not harm anyone.	1.11	13.8	0.59	9
23. Space exploration must continue so we can learn if there is life on other planets.	1.26	14.4	0.59	35
24. The space program must be continued so we do not lose the capabilities we have developed for spaceflight and the rocket know-how we have invested so much to gain.	1.47	17.3	0.62	16
25. The success of the U.S. space program increases our prestige in the world, demonstrates the value of democracy, and renews American national pride.	1.39	18.8	0.55	27
26. The space program must be continued so that its highly trained technical manpower will not be wasted in unemployment.	0.88	8.0	0.60	17
27. Space is an important arena for international competition, and if	1.20	16.1	0.64	1

	Mean Rating	Percent High Rating	Strongest Link	
			Corre-lation	With Item
we do not keep our lead, the Russians will gain an advantage over us.				
28. The space program encourages young people to choose careers in science and technology, and is itself a good training ground for scientists and engineers.	1.71	21.4	0.51	16
29. Space progress will provide new, rewarding jobs for many people.	1.71	22.8	0.60	17
30. Human societies have always needed to expand in order to remain healthy; space is the only direction left for such expansion.	1.00	12.3	0.60	33
31. Men and societies have always needed the challenges provided by a frontier; space is the new frontier.	1.48	17.8	0.61	42
32. Space travel will lead to the planting of human colonies on new worlds in space.	0.89	8.6	0.68	34
33. Our world has become too small for human civilization and for the human mind; we need the wide open spaces of the stars and planets to get away from the confines of our shrinking world.	0.62	5.4	0.60	30
34. Society has a chance for a completely fresh start in space; new social forms and exciting new styles of life can be created on other worlds.	0.77	6.3	0.68	32
35. Communication with intelligent beings from other planets would give us completely new perceptions of humanity, new art, philosophy, and science.	1.59	30.1	0.59	23

	Mean Rating	Percent High Rating	Strongest Link	
			Corre-lation	With Item
36. Human society on Earth needs the change and global renewal that space travel will bring.	0.75	8.2	0.59	40
37. Space can provide a focus for increasing international cooperation leading to world unity.	1.84	29.1	0.47	21
38. Spaceflight is necessary to ensure the survival of the human race against destruction by natural or man-made disaster.	0.94	11.9	0.57	34
39. I am in favor of the space program because I would very much like the experience of traveling into space myself.	1.69	12.2	0.52	40
40. Space exploration is an exciting adventure, valuable for the fun and excitement it provides.	0.67	5.0	0.61	42
41. We must explore space for the same reason people climb Mount Everest—because it's there.	0.76	6.3	0.55	42
42. We must explore space to satisfy our great curiosity; space exploration is an expression of man's natural curiosity.	1.37	14.4	0.61	31
43. Travel into space will teach us to love and respect our own planet, Earth.	1.16	16.4	0.48	29
44. Space can provide us with a goal and sense of purpose which mankind badly needs.	0.92	8.5	0.49	31
45. Spaceflight enlarges the mind and the spirit of man, so that his ideas become universal rather than Earth-bound.	1.38	16.1	0.48	31

	Mean Rating	Percent High Rating	Strongest Link	
			Corre-lation	With Item
46. Without spaceflight we would be trapped, closed-in, jailed on this planet.	0.51	4.5	0.55	33
47. Mankind needs to know it is not alone in the universe, both to gain humility and to lose the feeling of loneliness.	0.64	5.5	0.50	36
48. Space will be of value in ways we cannot yet imagine.	2.03	37.1	0.58	21
49. In entering the universe mankind fulfills its destiny—to mature as a species and perhaps come closer to God.	1.62	6.0	0.37	34

APPENDIX B:
SPACE GOALS FROM THE AUTUMN 1986 HARVARD SURVEY

These 125 space goals were rated by 894 Harvard students, as explained in Chapter 3. For each, they circled a number (0–6) to indicate how good a reason they thought it was for supporting the space program: from "0" (not a good reason) to "6" (an extremely good reason). The first column of the table shows the mean ratings on this 0 to 6 scale. The second column shows the percent of those responding who rated the goal a 5 or 6, thus describing it as an extremely good reason for supporting the space program. The third and fourth columns show the strongest correlation possessed by each goal with another and the number of that other goal in the list. The numbering of items is arbitrary, simply the arrangement that was used in the computer analysis, and the space goals were presented in five different, random orders in five editions of the actual questionnaire.

	Mean Rating	Percent High Rating	Strongest Link Corre-lation	With Item
1. Space stimulates the creative, human imagination.	2.79	20.5	0.58	3
2. If America doesn't advance into space, some other country will.	1.72	11.0	0.67	21
3. Space exploration fulfills the human need for adventure.	2.27	13.2	0.66	14
4. Space discovery has religious implications, leading us to God.	0.72	3.4	0.34	82

	Mean Rating	Percent High Rating	Strongest Link	
			Corre- lation	With Item
5. Space is psychologically satisfying.	1.60	7.4	0.59	14
6. Eventually, interstellar travel could be possible, taking people to distant stars.	2.96	25.1	0.58	64
7. Farms in space and advances in terrestrial agriculture aided by the space program could increase our food supply.	3.50	33.9	0.49	92
8. The space program could save the human race—by warding off a threat to the existence of our planet, for example.	2.36	16.0	0.47	111
9. The long-term, ultimate benefits of the space program could eventually be important.	4.24	52.9	0.54	113
10. We could understand the Earth's atmosphere and geology better, through comparisons with other planets.	3.92	38.6	0.57	88
11. Competition in space is a constructive outlet for nationalistic rivalries that otherwise would take the form of aggression and conflict.	1.78	7.0	0.41	85
12. New medicines could be manufactured in the zero gravity and vacuum of space.	4.37	53.0	0.68	79
13. Vacations and games in space could be entertaining.	1.43	8.5	0.57	52
14. Investigation of outer space satisfies human curiosity.	2.46	15.9	0.66	3
15. The space program contributes much to our technology.	4.11	45.9	0.58	33
16. Space travel makes us realize that Earth is a fragile, unique,				

	Mean Rating	Percent High Rating	Strongest Link	
			Corre-lation	With Item
unified world that deserves more respect and better care.	3.00	27.3	0.54	123
17. The worst criminals could be put in space prisons.	1.05	7.3	0.35	13
18. Space research helps us understand our place in the universe.	3.03	23.4	0.52	29
19. Satellites link all corners of the globe in a complete information and communication network.	4.24	49.9	0.53	75
20. Space is the new frontier.	2.59	20.8	0.59	101
21. The United States must develop space technology vigorously to keep up with the Soviet Union and other countries.	2.18	14.8	0.67	2
22. The space program is an educational tool, helping us learn from each other.	2.87	17.7	0.48	34
23. Observations from orbit help us find new sources of energy and minerals on the Earth.	3.94	39.1	0.51	68
24. We could establish manned space stations, communities in space, and space cities.	3.09	24.7	0.68	64
25. Space missions are exciting.	1.77	11.0	0.66	52
26. The space program increases national prestige, producing worldwide respect for America.	1.91	10.5	0.67	67
27. Jokes and cartoons about the space program can be amusing.	0.71	5.7	0.46	13
28. We could learn much from contact and communication with intelligent, extraterrestrial beings.	2.65	22.3	0.64	51
29. The space program gives us				

	Mean Rating	Percent High Rating	Strongest Link Corre-lation	Strongest Link With Item
new perspectives on ourselves and our world.	2.86	20.1	0.56	1
30. Reconnaissance satellites help prevent war and nuclear attack.	2.77	22.0	0.52	109
31. The space program contributes to the advancement of science.	4.46	54.9	0.59	38
32. Astronauts are heroes and role models for young adults.	1.94	8.1	0.56	81
33. Technological spin-offs (advancements developed for the space program, then applied to other fields) improve every-day life.	3.98	41.9	0.58	15
34. Joint space projects between nations improve international cooperation.	3.44	30.4	0.64	56
35. The space program stimulates the economy and has direct economic benefits.	2.81	17.1	0.52	106
36. The exploration of space is an unselfish quest that could benefit all mankind.	2.78	20.4	0.48	48
37. There are great military applications of space.	1.67	10.9	0.72	109
38. Space research tests our scientific theories and promises conceptual breakthroughs.	4.18	47.3	0.59	31
39. We could learn more about life by experimenting with life processes in the different conditions of space and other planets.	3.46	27.6	0.45	38
40. We could find new mineral resources on the Moon, Mars, or the asteroids.	3.59	32.4	0.66	121

	Mean Rating	Percent High Rating	Strongest Link Corre-lation	Strongest Link With Item
41. We gain knowledge, and it is good to have knowledge for its own sake.	3.30	31.7	0.55	77
42. The exploration of space lifts morale and instills a sense of hope and optimism.	2.54	14.4	0.65	107
43. We could discover our own origins, learning about the history of the universe and Earth.	3.24	28.2	0.57	76
44. Meteorology satellites are great aids for predicting the weather and understanding atmospheric patterns.	4.25	46.2	0.56	68
45. The Earth is too small for us, so we must expand off this planet.	2.11	12.2	0.65	74
46. Humans should spread life to other planets.	1.75	9.3	0.63	45
47. AIDS victims could live safely in the complete isolation from infection that can be assured in space.	1.02	5.7	0.32	17
48. The common cause of space exploration unites the peoples of the world and could eventually create a world community.	2.71	20.5	0.63	34
49. Many experiments can be done best in the environment of space.	4.20	49.2	0.56	98
50. Humans have an innate need to search and discover.	2.92	21.8	0.63	14
51. We might find alien forms of life on other planets.	2.64	19.7	0.64	28
52. Space travel is fun.	1.67	13.9	0.66	25
53. Space could improve life on Earth, providing benefits to mankind.	3.85	40.5	0.49	89

	Mean Rating	Percent High Rating	Strongest Link	
			Corre-lation	With Item
54. Our ability to travel on Earth could be improved—with high-speed aircraft, for example.	3.28	24.7	0.46	33
55. The space program channels government spending away from destructive weapons, and it is better than wasting money on the military.	2.44	19.9	0.46	48
56. Space promotes cooperation between the United States and the Soviet Union, working together for a common goal.	3.06	25.3	0.64	34
57. Space gives people something to dream about.	2.06	13.5	0.61	61
58. Space probes increase our knowledge of space, planets, comets, and the entire solar system.	4.31	51.2	0.56	122
59. We could preserve Earth's environment by moving the most polluting industries into space.	2.53	19.8	0.59	73
60. The space program produces better computers, calculators, and electronics.	3.22	26.8	0.56	33
61. The beauty of space creates a sense of wonder.	2.15	14.8	0.64	82
62. From space, we could find new ways to control pollution and clean up our environment.	3.37	27.2	0.45	112
63. Solar power stations in orbit could provide clean, limitless energy to the Earth.	4.24	51.0	0.52	72
64. We could colonize the Moon, Mars, and other satellites or planets of our solar system.	2.50	16.1	0.68	24
65. We could gain greater				

	Mean Rating	Percent High Rating	Strongest Link	
			Corre-lation	With Item
understanding of the world we live in.	3.46	28.2	0.53	100
66. We have the capability to explore space now, and we must not waste this opportunity.	2.71	18.8	0.54	101
67. The space program builds national pride.	2.30	13.1	0.67	26
68. Satellites are useful in surveying and mapping the Earth.	3.89	36.2	0.63	75
69. Sex in the zero gravity of space could be quite an experience.	2.60	35.2	0.36	13
70. The space program could help us control the weather, bringing rain to drought-stricken areas.	3.08	23.5	0.41	7
71. Space offers new challenges, and civilization would stagnate without challenges.	2.51	16.7	0.60	124
72. New fuels found in space or the development of fusion power in space could help solve the Earth's energy problem.	4.03	45.0	0.55	40
73. The Moon or the sun could be used for safe disposal of toxic materials and nuclear wastes.	2.65	21.8	0.59	59
74. Space offers room for the expansion of the human species.	2.74	18.8	0.65	92
75. Satellites are an important component in navigation systems.	3.76	33.6	0.63	68
76. Through the space program we could learn the origin of life.	2.63	18.8	0.57	43
77. We should explore the unknown.	3.05	26.4	0.60	50
78. Space research benefits physics — in studies of the nature of matter, for example.	4.08	41.7	0.56	100
79. Medical research performed in space could benefit human health.	4.35	53.8	0.68	12

	Mean Rating	Percent High Rating	Strongest Link	
			Corre- lation	With Item
80. New experiences and perspectives gained in space inspire art, music, and literature.	2.07	10.0	0.56	61
81. Spaceflight reaffirms faith in man's abilities.	2.12	12.2	0.71	125
82. There are wonderful sights in space, such as Earth and Saturn, that we can see in pictures.	1.68	8.4	0.64	61
83. Our future ultimately lies in space.	2.70	22.8	0.57	74
84. Living in space could teach us the value of conservation, efficiency, and environmental responsibility.	2.95	21.8	0.49	16
85. The space program generates national unity, encouraging cooperation between numerous sectors of society.	2.67	16.4	0.58	67
86. The space program employs many engineers and scientists who otherwise would not be able to utilize their talents.	2.27	13.0	0.54	90
87. The space program represents the best traditions of Western Civilization.	1.46	5.4	0.53	125
88. Satellite photography of the Earth contributes to geology, oceanography, and archaeology.	4.00	38.4	0.62	68
89. Space research provides valuable, practical information.	4.09	44.2	0.55	15
90. The space program provides jobs for thousands of people.	3.01	20.0	0.54	86
91. I want to travel in space.	2.32	24.6	0.58	52
92. Space settlements could ease the growing problem of overpopulation.	2.61	16.8	0.65	74

	Mean Rating	Percent High Rating	Strongest Link	
			Corre-lation	With Item
93. We could gain knowledge about ourselves.	2.64	16.8	0.52	65
94. Limitless opportunities could be found in space.	2.93	24.3	0.55	83
95. The space program inspires young people to study the sciences.	3.23	26.0	0.48	105
96. A space-based anti-missile, system, part of the Strategic Defense Initiative, could reduce the danger of war and nuclear annihilation.	1.82	15.0	0.69	37
97. The space program has great benefits for industry.	3.32	27.6	0.65	106
98. In the weightlessness and vacuum of space, we could manufacture new and better alloys, crystals, chemicals, and machine parts.	4.18	47.8	0.60	12
99. Some medical problems could be treated more effectively in the weightlessness of space.	3.87	38.6	0.68	79
100. We could gain a better understanding of the universe as a whole and how it functions.	3.97	41.2	0.56	78
101. We must broaden our horizons.	2.46	14.3	0.59	20
102. The space program contributes to world peace.	2.42	16.0	0.57	48
103. I like watching rockets take off and enjoy space probes.	1.17	7.6	0.60	25
104. Space triumphs give us justified pride in our achievements.	2.14	11.0	0.64	81
105. The space program encourages people to make achievements and solve problems.	2.85	19.1	0.53	81

	Mean Rating	Percent High Rating	Strongest Link	
			Corre-lation	With Item
106. Space has great commercial applications and many opportunities for business.	2.93	19.9	0.65	97
107. Spaceflight is a noble endeavor, expressing the hopes and aspirations of humankind.	2.43	15.8	0.69	125
108. We should go into space for the same reason people climb Mt. Everest—because it's there.	2.24	19.2	0.56	120
109. The space program contributes to our defense.	2.25	14.4	0.72	37
110. We could find new worlds we can live on or transform a planet to make it habitable.	2.66	17.1	0.66	64
111. We need an alternate home planet in case the Earth is destroyed by a natural catastrophe or nuclear war.	1.80	11.5	0.54	45
112. Through space research we learn the true extent of devastation that nuclear war would bring: nuclear winter.	2.28	14.3	0.45	62
113. Space could offer many unexpected benefits we cannot now foresee.	4.05	46.3	0.54	9
114. In space, we could create new cultures, lifestyles, and forms of society.	1.88	9.6	0.47	94
115. Progress in space is part of the advancement of mankind.	3.07	27.3	0.61	107
116. The space program allows people to think beyond the triviality of Earth-bound conflicts and concerns.	2.43	15.7	0.55	42
117. We must be able to launch, retrieve, and repair satellites.	3.72	36.7	0.50	75

	Mean Rating	Percent High Rating	Strongest Link	
			Corre-lation	With Item
118. Mankind is bound to venture outside Earth and needs to know as much as possible in advance.	2.77	20.2	0.49	74
119. Communication satellites improve television transmissions.	2.96	22.7	0.50	75
120. We should boldly go where no man or woman has gone before.	2.35	22.0	0.59	77
121. We could use raw materials from the moon and planets when natural resources are depleted on Earth.	3.49	32.2	0.66	40
122. An orbiting space telescope could give astronomers a much better view of the stars.	3.90	40.0	0.56	58
123. In space, we see how small our world is and thus learn humility.	1.83	11.6	0.54	16
124. The space program provides a goal and a feeling of long-term purpose for humanity.	2.45	13.6	0.62	125
125. Space exploration is a human struggle, expressing the unconquerable human spirit.	2.09	11.2	0.71	81

BIBLIOGRAPHY

Allaby, Michael, and James Lovelock. 1984. *The Greening of Mars*. New York: St. Martin's Press.

Almond, Gabriel A. 1960. "Public Opinion and the Development of Space Technology." *Public Opinion Quarterly* 24: 553–572.

Bainbridge, William Sims. 1976. *The Spaceflight Revolution*. New York: Wiley-Interscience.

———. 1978a. "Public Support for the Space Program," *Astronautics and Aeronautics* 16 (6) 60–61, 76. reprinted in *Journal of Contemporary Business* 1987, 7 (3): 185–189.

———. 1978b. *Satan's Power*. Berkeley: University of California Press.

———. 1982a. "The Impact of Science Fiction on Attitudes toward Technology." In *Science Fiction and Space Futures*, ed. Eugene M. Emme, pp. 121–135. San Diego: American Astronautical Society.

———. 1982b. "Religions for a Galactic Civlization." Pp. 187–201 in *Science Fiction and Space Futures*, edited by Eugene M. Emme. San Diego: American Astronautical Society.

———. 1983. "Attitudes toward Instellar Communication." *Journal of the British Interplanetary Society* 36: 298–304.

————. 1984. "Computer Simulation of Cultural Drift: Limitations on Interstellar Colonisation." *Journal of the British Interplanetary Society* 37: 420–429.

————. 1985. *Experiments in Psychology.* Belmont, Calif.: Wadsworth Publishing Co.

————. 1986. *Dimensions of Science Fiction.* Cambridge, Mass.: Harvard University Press.

————. 1987a. "Collective Behavior and Social Movements." In Rodney Stark, *Sociology,* Belmont, Calif.: Wadsworth Publishing Co.

————. 1987b. *Sociology Laboratory.* Belmont, Calif.: Wadsworth Publishing Co.

————. 1989a. *Survey Research: A Computer-Assisted Introduction.* Belmont, Calif.: Wadsworth Publishing Co.

————. 1989b. *Survey Research: Instructor's Manual.* Belmont, Calif.: Wadsworth Publishing Co.

Bainbridge, William Sims, and Robert D. Crutchfield. 1983. "Sex Role Ideology and Delinquency," *Sociological Perspectives* 26: 253–274.

Bainbridge, William Sims, and Laurie Russell Hatch. 1982. "Women's Access to Elite Careers: In Search of a Religion Effect." *Journal for the Scientific Study of Religion* 21: 242–254.

Bainbridge, William Sims, and Rodney Stark. 1981a. "The 'Consciousness Reformation' Reconsidered." *Journal for the Scientific Study of Religion* 20: 1–16.

————. 1981b. "Friendship, Religion, and the Occult: A Network Study." *Review of Religious Research* 22: 313–327.

Bainbridge, William Sims and Richard Wyckoff. 1979. "American Enthusiasm for Spaceflight," *Analog* 99 (July): 59–72.

Bauer, Raymond A. 1960. "Executives Probe Space." *Harvard Business Review* (September–October): 6–14, 174–200.

———. 1969. *Second-Order Consequences*. Cambridge, Mass.: MIT Press.

Bell, Trudy E. 1980a. "American Space-Interest Groups." *Star and Sky* (September): 53–60.

———. 1980b. "The Grand Analogy: History of the Idea of Extraterrestrial Life." *Cosmic Search* 2: 2.

Bester, Alfred. 1956–1957. "Tiger! Tiger! (*The Stars My Destination*)." *Galaxy* 12 (October 1956): 8–58; 13 (November 1956): 88–143; (December 1956): 88–142; (January 1957): 98–142.

Block, Jack. 1965. *The Challenge of Response Sets*. New York: Appleton-Century-Crofts.

Bradbury, Ray. 1977. "Introduction." In T. A. Heppenheimer, *Colonies in Space*. Harrisburg, Penn.: Stackpole Books.

Browning, Robert. 1934. *The Shorter Poems of Robert Browning*. New York: Appleton-Century-Crofts.

Bonestell, Chesley, and Willy Ley. 1949. *The Conquest of Space*. New York: Viking Press.

Bova, Ben. 1981. *The High Road*. Boston: Houghton Mifflin.

Broad, William J. 1990. "A Search Goes On: Are We Alone in the Universe?" *New York Times* (February 6), pp. B5, B8.

Cantril, Hadley. 1966 [1940]. *The Invasion from Mars*. New York: Harper Books.

Carr, Michael H. 1981. *The Surface of Mars*. New Haven, Conn.: Yale University Press.

Cheston, T. Stephen, Charles M. Chafer, and Sallie Birket Chafer. 1984. *Social Sciences and Space Exploration*. Washington, D.C.: National Aeronautics and Space Administration.

Collins, Michael. 1988. "Mission to Mars." *National Geographic* 174, no. 5: 732–764.

Compton, W, David, and Charles D. Benson. 1983. *Living and Working in Space: A History of Skylab*. Washington, D.C.: National Aeronautics and Space Administration.

Cooper, Henry S. F. 1980. *The Search for Life on Mars*. New York: Holt, Rinehart and Winston.

Davis, James A., and Tom W. Smith. 1986. *General Social Surveys, 1972–1986: Cumulative Codebook*. Chicago: National Opinion Research Center.

Dick, Steven J. 1982. *Plurality of Worlds*. London: Cambridge University Press.

Dyson, Freeman. 1979. *Disturbing the Universe*. New York: Harper and Row.

Economist, The. 1986. "Questions Across the Atlantic." *The Economist* (December 13): 56.

Everitt, Brian. 1974. *Cluster Analysis*. London: Heinemann.

Ezell, Edward Clinton, and Linda Neuman Ezell. 1978. *The Partnership: A History of the Apollo-Soyuz Test Project*. Washington, D.C.: National Aeronautics and Space Administration.

———. 1984. *On Mars: Exploration of the Red Planet 1958–1978*. Washington, D.C.: National Aeronautics and Space Administration.

Festinger, Leon, H. W. Riecken, and Stanley Schachter. 1956. *When Prophecy Fails.* New York: Harper and Row.

Festinger, Leon, Stanley Schachter, and Kurt Back. 1950. *Social Pressures in Informal Groups.* New York: Harper Books.

Foley, Maura Stephens. 1986. "The Public's View." *Newsweek* (February 10), p. 37.

Fores, Michael. 1969. "All One Culture—Or Three?" *New Scientist* (19 June), pp. 637–638.

Furash, Edward E. 1963. "Businessmen Review the Space Effort." *Harvard Business Review* (September–October): 14–32, 173–190.

Gallup, George. 1981. "Teen-Agers Slimly Favor U.S. Space Program." *Insight—National Space Institute* 5 (January–February): 7 (from the Associated Press).

Gallup, George H. (ed.). 1972. *The Gallup Poll: Public Opinion 1935–1971.* New York: Random House.

Gallup Opinion Index. 1966. "First Moon Flight" (July).

———. 1969. "Funds for Space Research" (March): 17.

———. 1974. (January): 20.

———. 1977. *Religion in America 1977–78.* Princeton, N.J.: American Institute of Public Opinion.

Glock, Charles Y., and Rodney Stark. 1965. *Religion and Society in Tension.* Chicago: Rand McNally.

———. 1966. *Christian Beliefs and Anti-Semitism.* New York: Harper and Row.

Goethe, Johann Wolfgang von. 1965. *Faust: Part One and Part*

Two, trans. Charles E. Passage. Indianapolis: Bobbs-Merrill Publishing Co.

Goodwin, Harold Leland. 1965. *The Images of Space*. New York: Holt, Rinehart and Winston.

Gusfield, Joseph R. 1963. *Symbolic Crusade*. Urbana: University of Illinois Press.

Gutting, Gary (ed.). 1980. *Paradigms and Revolutions*. Notre Dame, Ind.: Notre Dame University Press.

Hansen, William P., and Fred L. Israel (eds.). 1972. *The Gallup Poll: Public Opinion 1935–1971*. New York: Random House.

Harvard Gazette. 1983. "Harvard Telescope to Search the Stars for Intelligent Life." *Harvard Gazette* (January 14), pp. 1, 8.

Hofstadter, Richard. 1963. *Anti-Intellectualism in American Life*. New York: Vintage Books.

Hunt, John. 1954. *The Conquest of Everest*. New York: E. P. Dutton.

James, William. 1911. "The Moral Equivalent of War." In *Memories and Studies*, pp. 267–296. New York: David McKay Co.

Jewkes, John, David Sawers, and Richard Stillerman. 1969. *The Sources of Invention*. New York: W. W. Norton and Co.

Joëls, Kerry Mark. 1985. *The Mars One Crew Manual*. New York: Balantine Books.

Johnson, Richard D., and Charles Holbrow (eds.). 1977. *Space Settlements: A Design Study*. Washington, D.C.: National Aeronautics and Space Administration.

Kash, Don E. 1967. *The Politics of Space Cooperation*. Lafayette, Ind.: Purdue University Studies.

Klass, Philip J. 1971. *Secret Sentries in Space.* New York: Random House.

Kuhn, Thomas S. 1959. *The Copernican Revolution.* New York: Vintage Books.

———. 1962. *The Structure of Scientific Revolutions.* Chicago: University of Chicago Press.

Lewis, C. S. 1965. *Out of the Silent Planet.* New York: Macmillan Publishing Co.

Ley, Willy. 1969. *Rockets, Missiles, and Men in Space.* New York: Signet Books.

Ley, Willy, and Wernher von Braun. 1956. *The Exploration of Mars.* New York: Viking Press.

Lipset, Seymour Martin, and Earl Raab. 1970. *The Politics of Unreason.* New York: Harper and Row.

Logsdon, John M. 1970. *The Decision to Go to the Moon.* Cambridge, Mass.: MIT Press.

Lunan, Duncan. 1974. *Interstellar Contact.* Chicago: Henry Regnery Co.

———. 1979. *New Worlds for Old.* Newton Abbot, England: Westbridge.

———. 1983. *Man and the Planets.* Bath, England: Ashgrove.

Lurie, Alison. 1967. *Imaginary Friends,* New York: Avon.

MacGowan, Roger. A., and Frederick. I. Ordway, III. 1966. *Intelligence in the Universe.* Englewood Cliffs, N.J.: Prentice-Hall.

MacLeish, Archibald. 1978. *Riders on the Earth.* Boston: Houghton Mifflin.

Malinowski, Bronislaw. 1948. *Magic, Science and Religion.* New York: Delacorte Press.

Malthus, Thomas R. 1888. *An Essay on the Principle of Population.* London: Reeves and Turner.

McDougall, Walter A. 1985. *. . . the Heavens and the Earth.* New York: Basic Books.

Mendell, Wendell W. (ed.). 1985. *Lunar Bases and Space Activities of the 21st Century.* Houston: Lunar and Planetary Institute.

Merton, Robert K. 1970. *Science, Technology and Society in Seventeenth-Century England.* New York: Harper and Row.

Michael, Donald N. 1960. "The Beginning of the Space Age and American Public Opinion." *Public Opinion Quarterly* 24: 573–582.

Michaud, Michael A. G. 1986. *Reaching for the High Frontier,* New York: Praeger.

Miller, Ron, and Frederick C. Durant. 1983. *Worlds Beyond: The Art of Chesley Bonestell.* Norfolk, Va.: Donning Co.

Molton, P. M. 1978. "On the Likelihood of a Human Interstellar Civilization." *Journal of the British Interplanetary Society* 31: 203.

Morrison, David. 1982. *Voyages to Saturn.* Washington, D.C.: National Aeronautics and Space Administration.

Murray, Bruce, Samuel Gulkis, and Robert E. Edelson. 1978. "Extraterrestrial Intelligence: An Observational Approach." *Science* 199: 485–492.

Murray, Bruce, Michael C. Malin, and Ronald Greeley. 1981. *Earthlike Planets.* San Francisco: W. H. Freeman and Co.

National Commission on Space. 1986. *Pioneering the Space Frontier*, New York: Bantam.

New York Times. 1985. "Survey Indicates a Sharp Contrast in Views on Missile Defense Plan." *New York Times* (January 10): p. A10.

New York Times. 1986. "Physicists Call Missile Shield an Error, Poll Finds." *New York Times* (March 23): p. 26.

Oberg, James Edward. 1981. *New Earths: Transforming Other Planets for Humanity*. Harrisburg, Penn.: Stackpole Books.

Ogburn, William Fielding. 1922. *Social Change*. New York: Huebsch.

O'Leary, Brian. 1983. *Project Space Station*. Harrisburg, Penn.: Stackpole.

O'Neill, Gerard K. 1977. *The High Frontier*. New York: Bantam Books.

Ordway, Frederick I., Carsbie C. Adams, and Mitchell R. Sharpe. 1971. *Dividends from Space*. New York: Thomas Y. Crowell.

Ouchi, William G. 1981. *Theory Z: How American Business Can Meet the Japanese Challenge*. New York: Avon Books.

Pareto, Vilfredo. 1935. *Non-Logical Conduct*. Volume I of *The Mind and Society*. New York: Harcourt Brace.

Riabchikov, Evgeny. 1971. *Russians in Space*. Garden City, N.Y.: Doubleday Books.

Rosengren, Karl Erik, Peter Arvidson, and Dahn Sturesson. 1975. "The Barsebäck 'Panic': A Radio Programme as a Negative Summary Event." *Acta Sociologica* 18: 147–162.

Rynin, N. A. 1931. *Interplanetary Flight and Communication*. Vol.

I, No. 3. Jerusalem: Israel Program for Scientific Translations, reprinted 1971.

Sagan, Carl. 1983. "Nuclear War and Climatic Catastrophe." *Foreign Affairs* 62: pp. 257–292.

———. (ed.). 1973. *Communication with Extraterrestrial Intelligence.* Cambridge, Mass.: MIT Press.

Sanger, Eugen. 1958. *Raumfahrt—Technische Überwindung des Krieges.* Hamburg, Germany: Rowohlt.

Sapolsky, Harvey M. 1972. *The Polaris System Development.* Cambridge, Mass.: Harvard University Press.

Schmookler, Jacob. 1966. *Invention and Economic Growth.* Cambridge, Mass.: Harvard University Press.

Schutz, Alfred. 1967. *The Phenomenology of the Social World.* Evanston, Ill.: Northwestern University Press.

Shapley, Harlow. 1963. *The View from a Distant Star.* New York: Basic Books.

Simon, Leslie E. 1971. *Secret Weapons of the Third Reich: German Research in World War II.* Old Greenwich, Conn.: WE.

Singer. C. E. 1982. "Galactic Extraterrestrial Intelligence," *Journal of the British Interplanetary Society* 35: 99.

Skelly, Florence. 1986. "The Stanford/Harvard Survey." *Harvard Magazine* 88 (March–April): 21–27.

Snow, C. P. 1969. *The Two Cultures.* London: Cambridge University Press.

Spengler, Oswald. 1926. *The Decline of the West*, trans. Charles Francis Atkinson. New York: Alfred Knopf and Co.

Stark, Rodney, and William Sims Bainbridge. 1985. *The Future of Religion*. Berkeley: University of California Press.

———. 1987. *A Theory of Religion*. New York: Peter Lang.

Stark, Rodney, and Charles Y. Glock. 1968. *American Piety*. Berkeley: University of California Press.

Stephenson, D. G. 1977. "Factors Limiting the Interaction between Twentieth Century Man and Interstellar Civilization," *Journal of the British Interplanetary Society* 30: 105.

Stevens, C. J. 1969. *Astrotheology*. Techny, Ill.: Divine Word.

Stoiko, Michael. 1970. *Soviet Rocketry*. New York: Henry Holt.

Strickland, Donald A. 1965. "Physicists' Views of Space Politics." *Public Opinion Quarterly* (Summer): 223–235.

Taviss, Irene. 1972. "A Survey of Popular Attitudes toward Technology." *Technology and Culture* 13: 606–621.

Tennyson, Alfred Lord. 1910. *The Poetic and Dramatic Works of Alfred Lord Tennyson*. Boston: Houghton Mifflin.

Tokaev, G. A. 1951. *Stalin Means War*. London: Weidenfeld.

Tough, Allen. 1986. "What Role Will Extraterrestrials Play in Humanity's Future?" *Journal of the British Interplanetary Society* 39: 491–498.

Turco, R. P., O. B. Toon, T. P. Ackerman, J. B. Pollack, and Carl Sagan. 1983. "Nuclear Winter: Global Consequences of Multiple Nuclear Explosions." *Science* 222: 1283–1292.

Turner, Frederick Jackson. 1920. *The Frontier in American History*. New York: Henry Holt.

Van Dyke, Vernon. 1964. *Pride and Power—The Rationale of the Space Program*. Urbana: University of Illinois Press.

von Braun, Wernher. 1952. "Prelude to Space Travel." In *Across the Space Frontier*, ed. Cornelius Ryan, pp. 12–70. New York: Viking Press.

————. 1953. *The Mars Project*. Urbana: University of Illinois Press.

von Hoerner, Sebastian. 1975. "Population Explosion and Interstellar Expansion." *Journal of the British Interplanetary Society* 28: 691–712.

Weber, Max. 1958. *The Protestant Ethic and the Spirit of Capitalism*. New York: Scribner's.

White, Frank. 1987. *The Overview Effect*. Boston: Houghton Mifflin.

White, Harrison C., Scott A. Boorman, and Ronald L. Breiger. 1976. "Social Structure from Multiple Networks." *American Journal of Sociology* 81: 730–780.

White, Leslie A. 1959. *The Evolution of Culture*. New York: McGraw-Hill Book Co.

Winter, Frank H. 1983. *Prelude to the Space Age*. Washington, D.C.: Smithsonian Institution.

York, Herbert F. 1985. "Nuclear Deterrence and the Military Uses of Space." *Daedalus* 114, no. 2, pp. 17–32.

INDEX

medicines, 69, 94–96, 105, 236
Mendell, Wendell W. 182
Merton, Robert K., 2
meteorology satellites, 35, 39, 52, 69, 71, 228, 239; international, 125; military 161. *See also* weather
Michael, Donald N., 148
Michaud, Michael, 33
military goals, 79, 81, 141–162, 225–226; and male support for spaceflight, 11; applications, 73, 227, 238; control of space, 18–19; factor in Seattle survey, 42–43; factor in University of Washington survey, 53–54; space station, 181; waste of money, 122–124, 240. *See also* defense, reconnaissance satellites, Strategic Defense Initiative
Miller, Ron, 128
mind, 46, 52, 231–232
mineral resources, 105, 176–177, 184, 237–238
mining, 22, 177; asteroids, 27
Molton, P. M., 194
Moon (Earth's), 22–27, 105, 162; colonization, 173, 175, 182, 191, 240; compared to Mars, 184; disposal of wastes, 178, 241; landing, 10, 120; prediction of trip, 9; resources, 177, 238
moral equivalent of war, 162
morale, 111, 113, 115, 239
Morrison, David, 183
Mount Everest, 46, 117–118, 120, 232, 244
Murray, Bruce, 183
music, 117, 128–129, 205, 242
Mutual Assured Destruction (MAD), 151
Mylar, 66
mystery, 127

National Aeronautics and Space Administration (NASA), 49, 60–61, 89–90, 92, 109, 120, 127, 130,

136, 179; future, 225; public support, 206, 217; safety record, 182
National Commission on Space, 29, 220
national unity, 128, 130, 147, 242
nationalism, 127–130, 145–148, 186
navigation satellites, 39, 84, 86–87, 89, 228, 241
Neptune, 25
nerve gas, 4
New England Science Fiction Association, 32
New York Times, 148
new worlds, 164, 173, 175, 193, 231, 244
Newsweek magazine, 16, 195
noble endeavor, 73, 112–113, 115, 187, 244
nomenclature, 84
Normandy Invasion, 161
Norway public opinion, 10, 147
nuclear power, 209. *See also* fusion
nuclear rockets, 23
nuclear waste, 27, 42, 177–178, 241
nuclear weapons, 4, 7, 144; disposal of, 28
nuclear winter, 88, 105–106, 168, 178–179, 244

O'Leary, 179
O'Neill, Gerard K., 27, 175
Oberg, James Edward, 176, 189
oceanography, 25, 70, 84, 86–88, 242
odyssey, 139
Ogburn, William Fielding, 2
Oort Cloud, 25
open ended survey items, 21, 55
opinion polls, 8–17; national 61. *See also* Gallup poll, General Social Survey
opportunity, 134; business, 91; limitless, 103, 165, 167, 243; to explore space, 131, 133, 241
orbital transfer vehicles, 178
order of goals, 85–86

Ordway, Frederick I., 108–109, 193
origin goals, 80–81, 106–108, 239, 241
Ouchi, William G., 109
overpopulation, 43, 163–165, 169–170, 173, 187–188, 193, 200, 230, 242
overview effect, 138

panel of judges, 62
paradigm, scientific, 3, 100
Pareto, Vilfredo, 111
patriotism, 129
peace, 122, 124, 145–146, 148, 243
Pentagon, 159
perspectives (new), 112, 117–118, 121, 128–129, 138, 204–205, 238, 242
PERT. *See* Program Evaluation and Review Technique
physicists; employment, 136; opinions, 17
physics, 70, 84, 94–95, 100–102, 210, 241
pilot study, 32
pioneer spirit, 187
plate tectonics, 26, 88, 184
Pluto, 25
poetry, 129, 139
Polaris missile, 109
politics; and extraterrestrials, 212; conservative versus liberal, 42, 154–160; instability, 223; justifications for spaceflight, 19
Pollack, J. B., 106
pollution, 80, 104–106, 168, 176–179, 240; electric power, 39, 228; reduced by space manufacturing, 24, 44
power, 147. *See also* elite
prediction of robot bombs, 9
prestige, 19, 42, 128–130, 147–148, 160, 230, 237
pride, 112–113, 116, 128–129, 134, 146–147, 160, 230, 241, 243
Program Evaluation and Review Technique (PERT), 109, 130
progress, 52, 101, 113, 115, 133–134, 152, 186, 244

propaganda, 125, 147
Protestantism, 111, 212–217
psychological justifications, 19, 117–119, 236
purpose, 111, 113–115, 232, 245

quest, 122, 238
questionnaire construction, 31–32

Raab, Earl, 217
racism, 220
radio; CETI, 207; telescope, 23, 197
random sample polls, 13, 15, 21, 36
raw materials, 43, 52, 105, 171, 176–177, 228, 245
Reagan, Ronald, 149, 157–158
reconnaissance satellites, 42, 53, 142–144, 160, 227, 238; manned, 158
Redstone missile, 158
religion, 20, 73; and CETI, 204, 212–218; justifications for spaceflight, 19; sociology, 59, 220; support for spaceflight, 213–214
renewal, 46, 232
repairing satellites, 86–89, 244
Republican Party, 154, 158
resources, 43, 53, 105, 176–177, 184–185, 228, 245
respect for Earth, 46, 131–132, 232, 237
response bias, 75
responsibility, 130–131
revolution; scientific, 3; spaceflight, 7
Rhea, 182
Riabchikov, Evgeny, 5
Riecken, H. W., 195
rockets; know-how, 41, 230; liquid-fuel versus solid-fuel, 4, 6
Rosengren, Karl Erik, 195
Russell, Bertrand, 49
Rynin, N. A., 170

Sagan, Carl, 106, 150, 185, 193
Salyut space station, 181